A NONLINEAR TIME SERIES WORKSHOP:

A Toolkit for Detecting and Identifying Nonlinear Serial Dependence

Dynamic Modeling and Econometrics in Economics and Finance

VOLUME 2

A NONLINEAR TIME SERIES WORKSHOP:

A Toolkit for Detecting and Identifying Nonlinear Serial Dependence

by

Douglas M. Patterson
Department of Finance, Insurance, and Business Law
Virginia Polytechnic Institute and State University
Blacksburg, Virginia

Richard A. Ashley
Department of Economics
Virginia Polytechnic Institute of Technology
Blacksburg, Virginia

KLUWER ACADEMIC PUBLISHERS
Boston / Dordrecht / London

Distributors for North, Central and South America:
Kluwer Academic Publishers
101 Philip Drive
Assinippi Park
Norwell, Massachusetts 02061 USA
Telephone (781) 871-6600
Fax (781) 871-6528
E-Mail <kluwer@wkap.com>

Distributors for all other countries:
Kluwer Academic Publishers Group
Distribution Centre
Post Office Box 322
3300 AH Dordrecht, THE NETHERLANDS
Telephone 31 78 6392 392
Fax 31 78 6546 474
E-Mail <services@wkap.nl>

Electronic Services <http://www.wkap.nl>

Library of Congress Cataloging-in-Publication Data

Patterson,Douglas M.(Douglas MacLennan),1945-
A nonlinear time series workshop : a toolkit for detecting and identifying nonlinear serial dependence / by Douglas M.Patterson, Richard A.Ashley.
p.cm.-- (Dynamic modeling and econometrics in economics and finance; v.2)
Includes bibliographical references and index.
ISBN 0-7923-8674-4(alk.paper)
1. Time-series analysis--Congresses.2.Economics,Mathematical--Congresses. I.Ashley,Richard A.(Richard Arthur),1950-II.Title.III.Series.
HA30.3 .P38 1999
330'.01'519232--dc21 99-046691

Printed on acid-free paper. Printed in the United States of America

Table of Contents

Preface

The analysis of what might be called "dynamic nonlinearity" in time series has its roots in the pioneering work of Brillinger (1965) – who first pointed out how the bispectrum and higher order polyspectra could, in principle, be used to test for nonlinear serial dependence – and in Subba Rao and Gabr (1980) and Hinich (1982) who each showed how Brillinger's insight could be translated into a statistical test. Hinich's test, because it takes advantage of the large sample statistical properties of the bispectral estimates became the first usable statistical test for nonlinear serial dependence. We are forever grateful to Mel Hinich for getting us involved at that time in this fascinating and fruitful endeavor.

With help from Mel (sometimes as a mentor, sometimes as a collaborator) we developed and applied this bispectral test in the ensuing period. The first application of the test was to daily stock returns {Hinich and Patterson (1982, 1985)} yielding the important discovery of substantial nonlinear serial dependence in returns, over and above the weak linear serial dependence that had been previously observed. The original manuscript met with resistance from finance journals, no doubt because finance academics were reluctant to recognize the importance of distinguishing between serial correlation and nonlinear serial dependence. In Ashley, Patterson and Hinich (1986) we examined the power and size of the test in finite samples. In Ashley and Patterson (1989) we examined the power of the test to detect the kinds of dynamic nonlinearity generated by typical macroeconomic models and obtained the first credible statistical evidence for dynamic nonlinearity in the generating mechanism for U.S. real output.

During this period the search for chaos – a particular form of deterministic nonlinear serial dependence – in macroeconomic time series received a good deal of attention in the time series community. Indeed, the well known BDS test was originally proposed as a test for chaos. In a widely circulated working paper {eventually published as Ashley and Patterson (1989)} we argued that the search for a purely deterministic generating mechanism for these data is a bit beside the point in that – since economic and financial time series are, in general, obviously stochastic – what is needed is nonlinear stochastic modeling. In that context, we suggested that the BDS test was more usefully viewed as another test for nonlinear serial dependence rather than as a test for deterministic chaos. We knew that our argument had prevailed when anonymous referee reports on the manuscript stopped asking for this section to be dropped because it was "wrong" and began instead to ask that it be dropped because the point was "widely accepted."

As a result of the application of a number of tests (Hinich Bispectral, BDS, and others) to economic and/or financial data, our focus of attention has shifted in recent years from the detection of nonlinearity to the tougher problem of identifying the form of the nonlinearity in a time series. The present volume reports on several strands in our efforts along these lines.

No project of this size flowers without the support of others. We greatly appreciate the ongoing encouragement (and computer codes) provided by Mel Hinich; and we would also like to thank Blake LeBaron and Houston Stokes for sharing computer codes with us. We also gratefully acknowledge the substantial contributions of Sumru Altug to the work summarized below in Chapter 10; she is our co-author in the journal article reporting on this work, which will appear in *Macroeconomic Dynamics* in late 1999. Finally, we thank our families for their forbearance during the course of this lengthy project.

References

Ashley, R. and D. Patterson (1989). "Linear Versus Nonlinear Macroeconomies" *International Economic Review 30*, 685-704.

Ashley, R., D. Patterson. M. and M. Hinich (1986). "A Diagnostic Test for Nonlinear Serial Dependence in Time Series Fitting Errors" *Journal of Time Series Analysis 7*, 165-78.

Brillinger, D. (1965) "An Introduction to Polyspectrum" *Annals of Mathematical Statistics 36*, 1351-74.

Hinich, M. (1982) "Testing for Gaussianity and Linearity of a Stationary Time Series" *Journal of Time Series Analysis 3*, 169-76.

Hinich, M. and D. Patterson (1982) "Evidence of Nonlinearity in Daily Stock Returns," manuscript, Department of Finance, Virginia Polytechnic Institute and State University.

_________________________(1985) "Evidence of Nonlinearity in Daily Stock Returns" *Journal of Business and Economic Statistics 3*, 69-77.

Subba Rao, T. and Gabr, M. (1980) "A Test for Linearity of Stationary Time Series Analysis" *Journal of Time Series Analysis 1*, 145-58.

1 NONLINEARITY IN STOCHASTIC PROCESSES: WHAT IT IS AND WHY IT MATTERS

OVERVIEW

Our purpose in writing this book is to make available a "nonlinearity toolkit" which provides the time series community with convenient, consistent access to a selection of the best tools available for statistically detecting nonlinearity in the generating mechanism of a given time series. Heretofore, these tests have been scattered through the literature, with computer codes implementing a few of the tests being transmitted informally from one person to another. As a result, workers in this area have often had to choose between either using the particular test(s) for which they had code or "reinventing the wheel" and writing their own computer codes. Consequently, it is not surprising that little testing for nonlinearity is done and that the testing which *is* done more often than not employs a rather arbitrarily selected and extremely narrow range of tests. The comparability, replicatability, and overall quality of the results have thereby suffered.

Our toolkit provides another benefit as well. The nonlinear identification problem – using the sample data to specify the form of the nonlinear process which generated it – has proven to be tremendously more difficult than the analogous identification problem for linear processes. Indeed, it is largely for want of an adequate solution to this problem that nonlinear statistical modeling is still in its infancy. But since our toolkit makes it straightforward to apply a battery of nonlinearity tests to the data, one can sensibly consider the possibility that the resulting *pattern* of test results may be helpful in nonlinear model identification. Indeed, the results we report on in the chapters below indicate that such patterns *can* convey useful identification information. (In particular, see Chapters 5, 9, and 10 below for examples.) We lay no claim to having solved the nonlinear model identification problem here, but the modest progress our approach provides is very welcome.

The remainder of this chapter addresses two issues. The first is fundamental: what is actually meant by nonlinearity in the stochastic process generating a time series? Constructively addressing this issue requires a brief review of the theory of stochastic processes. Since many sources on this topic already exist, our intent here is to provide an accessible overview which nevertheless goes deep enough in a few areas to give the reader a solid foundation for the material presented here. In fact, most of this book can be read and used without that solid foundation; in the long run, however, investing in that foundation will yield a large return. We see this section as a brief refresher for those who have already invested and as a uniquely accessible first step for those who have not.

The second issue taken up in this chapter is primarily motivational: why is the detection and modeling of nonlinearity in time series generating mechanisms so important? We argue below that a failure to recognize and deal with the presence of nonlinearity in the generating mechanism of a time series can often lead to poorly behaved parameter estimates and to models which miss important serial dependencies altogether.

STOCHASTIC PROCESSES: A BRIEF OVERVIEW

Defining Stochastic Processes

A stochastic process is a random experiment that evolves over time. A random experiment involves outcomes, ω, whose various objects belong to the sample space Ω (the certain event). To every ω the experiment assigns, according to some rule, a number.

$$x(\omega)$$

This number could be a "head" or a "tail" in a coin flipping experiment, or the price of oranges, or any other numerical quantity that is of interest to us in observing the experiment. The relationship between the elements ω of the set Ω and certain numbers defines a random number.

A stochastic process (or random function) is defined as a family of random variables $\{x_t; t \in T\}$. The subscript t is a parameter which indexes the random variable x, and belongs to the set T, the *index set*. In this book we will interpret t as time and we shall entertain two different index sets. When $T = \{...,-1,0,1...\}$ or $T = \{0,1,2,...\}$ the stochastic process is called a *discrete time* process. When $T = \{t; 0 \leq t < \infty\}$ or $T = \{t; -\infty < t < \infty\}$ the process is referred to as a *continuous time process.* Further, we will assume that the random variables x_t take values on the real line or in n - dimensional Euclidean space. A stochastic process is actually a function of two arguments $\{x_t(\omega); t \in T, \omega \in \Omega\}$ where Ω is the sample space, and the dependence on ω is usually suppressed when we write x_t. For fixed $t \in T$ $x_t(\cdot)$ is thus nothing more than a random variable and for fixed ω , $x_.(\omega)$ is a function of time which is called a *realization* of the process, a *trajectory*, a *path*, or a *sample function.* These sample functions can be thought of as elements of the space S which is usually called the *sample function space.*

What we shall call an *event* is a collection of outcomes on the sample space Ω (or the sample function space S) on which we have assigned probabilities. However, not all such collections can be considered as events. If Ω has infinitely many elements, it might be impossible to assign probabilities to all its subsets. In fact it is not possible to assign a probability measure to all subsets of Ω, only to a Borel field of subsets. A Borel field, or set, is obtained by a countable number of operations, starting from open sets, each operation consisting of taking unions, intersections, or complements.[1] The collection β of all Borel sets in r-dimensional Euclidean space is a σ-ring. With respect to applications, there is no serious limitation in the assignment of a probability measure for discrete time processes. The restriction is, however, a serious one for continuous time processes because, for example, the set

$$\{\omega; x_t(\omega) \le c \text{ for all } t \in (a, b)\}$$

is not a Borel field. Because of this, the measure of this set is not uniquely determined by the finite-dimensional distributions of the process.

A detailed exposition of the existence of a probability measure P on a collection of Borel fields, and of the related concept of a probability space $\{\Omega, \sigma, P\}$, is beyond the scope of this presentation. Suffice it to say that, for ordinary random variables whose sample function spaces are Euclidean spaces, probability measures can be assigned by ordinary distribution functions.

Another way to describe a stochastic process is to express x_t as a function of a known random process. As an example consider a discrete time process with $T = \{..., -1, 0, 1, ...\}$. Let the stochastic process $\{\eta_t, t \in T\}$ be a sequence of independent normal (0, 1) random variates. Next write the stochastic process $\{x_t, t \in T\}$ defined by

$$x_t = \eta_t + \theta_1\eta_{t-1} + \theta_2\eta_{t-2} + \ldots + \theta_q\eta_{t-q}, \qquad t \in T \tag{1.1}$$

This process is called a *moving average process* of order q. If the polynomial

$$z^q + \theta_1 z^{q-1} + \ldots + \theta_q = 0 \tag{1.2}$$

has roots inside the unit circle then the process generating x_t is said to be "invertible." In that case x_t can be expressed as an infinite order autoregressive or AR(∞) process. An AR process of order p is defined by

$$x_t + \varphi_1 x_{t-1} + \ldots + \varphi_p x_{t-p} = \eta_t. \qquad t \in T \tag{1.3}$$

which is called an *autoregressive process* of order p. Both the moving average and autoregressive processes are examples of *linear processes.*

More generally, a process is linear if it can be written as a linear aggregation of independent innovation terms

$$x_t = \sum_{j=0}^{\infty} \psi_j \eta_{t-j}, \qquad t \in T \tag{1.4}$$

where the ψ_j are parameters. It is not sufficient for the η_t innovations to be uncorrelated: they must be statistically independent. Note that (1.4) looks like the usual MA(∞) process, but an MA(∞) process merely has serially *uncorrelated* innovations.

The mean value of a process is defined by

$$\mu_t = E\{x_t\} = \int_{\Omega} x_t(\omega) P(d\omega) \qquad (1.5)$$

where $P(d\omega)$ is the probability measure and the symbol E denotes mathematical expectation. The probability assigned to ω is uniquely determined by a finite dimensional distribution function F, where

$$F(\xi_1, \xi_2, \ldots \xi_k, t_1, t_2, \ldots, t_k)$$

$$= P(x_{t_1} \le \xi_1, \xi_{t_2} \le \xi_2, \ldots, x_{t_k} \le \xi_k) \qquad (1.6)$$

Higher moments are defined in a similar fashion. For example, the covariance of x_s and x_t is given by

$$\mathrm{cov}(x_s, x_t) = E\{[x_s - \mu_s][x_t - \mu_t]\}$$

$$= \int_{\Omega} [x_s(\omega) - \mu_s][x_t(\omega) - \mu_t] P(d\omega), \qquad (1.7)$$

and the variance of x_t is the covariance of x_t with itself:

$$\mathrm{var}(x_t, x_t) = \sigma^2_{x_t} = E\{[x_t - \mu_t]^2\}$$

$$= \int_{\Omega} [x_t(\omega) - \mu_t]^2 P(d\omega). \qquad (1.8)$$

The definition of a stochastic process provided above is very general. In order for the theory to be manageable it is necessary to make a number of simplifying assumptions which make it possible to characterize the distribution of $x_{t_1}, x_{t_2}, \ldots, x_{t_x}$ in a simple manner.

STATIONARY PROCESSES

If the distribution of $x_{t_1}, x_{t_2}, \ldots, x_{t_x}$ is identical to the distribution of

$x_{t_1+\tau}, x_{t_2+\tau}, \ldots, x_{t_k+\tau}$ for all τ such that $t_i \in T$ and $t_i + \tau \in T$, $i = 1, 2, \ldots, k$, the process $\{x_t\}$ is called *stationary*. If only the first and second moments of the distributions are required to be the same, the stochastic process is said to be *weakly stationary.*

A stationary process is *ergodic* if the ensemble average equals the time-average of the sample functions:

$$E\{x_t\} = \int_{\Omega} x_t(\omega)P(d\omega) = \lim_{T\to\infty} \frac{1}{2T} \int_{-T}^{T} x_t(\omega)dt \tag{1.8}$$

for almost all ω. We note in passing that although one cannot prove that a process is ergodic, it is almost impossible to do inferential analysis of time series data without invoking an assumption of ergodicity.

NORMAL PROCESSES

Normal, or *Gaussian*, processes play a central role in time-series statistics. A stochastic process is normal if the joint distribution of $x_{t_1}, x_{t_2}, \ldots, x_{t_k}$ is normally distributed for every k and all $t_i \in T$, $i = 1, 2, \ldots, k$. A normal process is completely defined by its mean

$$\mu_i = E\{x_{t_i}\} \qquad i = 1,2,\ldots,k \tag{1.9}$$

and its covariance function

$$r_{ij} = \mathrm{cov}\{x_{t_i}, x_{t_j}\} = E\{[x_{t_i} - \mu_i][x_{t_j} - \mu_j]'\} \tag{1.10}$$

$$i,j = 1,2,\ldots,k$$

If we introduce the vector m and the matrix R:

$$m = \begin{bmatrix} \mu_1 \\ \mu_2 \\ \vdots \\ \mu_k \end{bmatrix} \qquad R = \begin{bmatrix} r_{11} & r_{12} \cdots & r_{1k} \\ r_{12} & r_{22} \cdots & r_{2k} \\ \vdots & & \\ r_{1k} & r_{2k} \cdots & r_{kk} \end{bmatrix} \tag{1.11}$$

where R is nonsingular, we can represent the joint distribution of $x_{t_1}, x_{t_2}, \ldots, x_{t_x}$ as the density function

$$f(\xi) = (2\pi)^{-k/2}(\det R)^{-1/2}\exp\left[-\frac{1}{2}(\xi - m)R^{-1}(\xi - m)\right] \tag{1.12}$$

It follows, therefore, that a normal process is completely specified if we know the mean and covariance functions for all $t_1, t_2, ..., t_k, t_k \in T$.

MARKOV PROCESSES

Let t_i and t be elements of the index set T such that $t_1 < t_2 < ... < t_k$. A stochastic process $\{x_t, t \in T\}$ is Markov if

$$P\left\{x_t \leqq \xi | x_{t_1}, x_{t_2}, ..., x_{t_k}\right\} = P\left\{x_{t_1} \leqq \xi | x_{t_k}\right\} \tag{1.13}$$

where $P\{\cdot|x_{t_k}\}$ denotes the conditional probability given x_t. If the *initial probability distribution*

$$F(\xi_1, t_1) = P\left\{x_{t_1} \leqq \xi_1\right\} \tag{1.14}$$

and the *transition probability distribution*

$$F(\xi_t, t|\xi_s, s) = P\left\{x_t \leq \xi_1 | x_s = \xi_s\right\} \tag{1.15}$$

are given, it then follows from Bayes' rule that the distribution function of the random variables $x_{t_1}, x_{t_2}, ..., x_{t}$ is given by the n-1 fold integral of the transition probability distribution[2]. A Markov process is therefore defined by two functions: the absolute probability distribution and the transition probabilities. Markov processes are frequently employed in modeling physical processes and economic relations. If a process is Markov it has the property that the future will evolve in a way that is statistically independent of the path taken to reach the current state of the process.

INDEPENDENT INCREMENTS PROCESSES

Consider a stochastic processes $\{x_t, t \in T\}$. Let $t_i \in T$ for i = 1, 2, ..., k, and $t_1 < t_2 < ... < t_k$. The process has *independent increments* if the random variables

$$x_{t_k} - x_{t_{k-1}},\ x_{t_{k-1}} - x_{t_{k-2}},\ ...\ x_{t_2} - x_{t_1},\ x_{t_1}$$

are mutually independent. If the variables are only uncorrelated then the process is said to be *uncorrelated,* or a process with *orthogonal increments.*

An independent increments process is specified by the distribution of the increment $x_t - x_s$ for arbitrary s and t; if the distribution of $x_t - x_s$ depends only on t - s, then the process has stationary increments.

CONVERGENCE CONCEPTS

What do we mean by the limit of a sequence of random variables such as $\{x_n(\omega), n = 1, 2, ...\}$? Below we will discuss the most common limit concepts employed in the study of stochastic processes.

Definition 1.1

The sequence $\{x_n(\omega)\}$ converges with *probability one* to the stochastic variable $x(\omega)$ if $x_n(\omega) \to x(\omega)$ for all w, except possibly for a set of ω-values having probability measure zero

$$P\left\{\omega;\ x_n(\omega) \to x(\omega)\right\} = 1 \tag{1.16}$$

Definition 1.2

The sequence $\{x_n(\omega)\}$converges to $x(\omega)$ in *probability* if for every $\varepsilon > 0$

$$\lim_{n \to \infty} P\left\{\omega;\ |x_n(\omega) - x(\omega)| \geq \varepsilon\right\} = 0 \tag{1.17}$$

Definition 1.3

The sequence $\{x_n(\omega)\}$ converges to $x(\omega)$ in the *mean square* sense if

$$\lim_{n \to \infty} E\left\{|x_n - x|^2\right\} = \lim_{n \to \infty} \int |x_n(\omega) - x_n|^2 P(d\omega) = 0 \tag{1.18}$$

Convergence with probability one implies convergence in probability; convergence in mean square implies convergence in probability[3]. The criteria for convergence with probability one are usually difficult to establish; we will use convergence in mean square here to keep the analysis tractable. We show below that there are simple criteria for differentiability and integrability in the mean square of second order processes; a second order process is one such that $E\{x_t^2\} < \infty$ for all $t \in T$.

WIENER PROCESS

The Wiener stochastic process (Brownian motion) is an important example of a stochastic process. Consider the following experiment: every T seconds toss a fair coin. A step up of height S is taken if heads show, and a step down is taken if tails show. Let x_t equal the position t seconds after the tossing is started. This is stochastic process known as a random walk. The paths have discontinuities at $t = nT$. A realization of the process for the sequence {h h t t t h h t} is shown in Figure 1-1.

Figure 1-1 Realization of a Random Walk

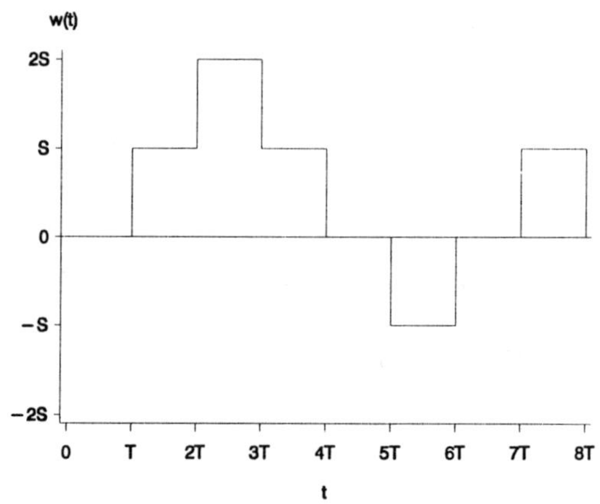

Suppose that after the first n tosses, k heads show. This implies that at $t = nT$, k steps up and n-k steps down have been taken; thus,

$$x_{nT} = kS - (n-k)S = (2k-n)S \tag{1.19}$$

Let $r = 2k-n$. It follows that r takes on the values

$$n, n-2, \ldots, -n,$$

and that

$$x_{nT} = rS, \tag{1.20}$$

and that

$$k = \frac{(r+n)}{2}. \tag{1.21}$$

The coin tosses constitute a sequence of Bernoulli trials. The probability of a heads showing on any trial is 1/2, independent of the outcomes of earlier trials. Therefore, the probability of k heads after n trials can be written:

$$\Pr\left\{x_{nT} = rS\right\} = \Pr\left\{\frac{r+n}{2} \text{ heads}\right\} = \binom{n}{\frac{r+n}{2}}\frac{1}{2^n} \tag{1.22}$$

Letting $x(i) = \pm S$ if the ith step is up or down, it follows directly that

$$\Pr\{x(i) = S\} = ½ \quad \text{and} \quad \Pr\{x(i) = -S\} = ½ \tag{1.23}$$

$$E\{x(i)\} = 0 \tag{1.24}$$

$$E\{x^2(i)\} = S^2 \tag{1.25}$$

Since the value of x_{nT} is

$$x_{nT} = x(1) + x(2) + \ldots + x(n) \tag{1.26}$$

where x(1), x(2), ... are independent, the variance of x_{nT} is nS^2.

From the DeMoivre-LaPlace limit theorem,[4]

$$\Pr\{k \text{ heads}\} \cong \frac{1}{\sqrt{2\pi npq}} e^{-(k-np)^2/2npq} \tag{1.27}$$

for n large. With p = q = ½, and k = (n+r)/2 this implies

$$\Pr\left\{x_{nT} = rS\right\} \cong \frac{1}{\sqrt{n\pi/2}} e^{-r^2/2n} \tag{1.28}$$

provided r is of order $\sqrt{n}$. It follows that

$$\Pr\{x_{nT} \le rS\} \cong \frac{1}{\sqrt{2\pi}} \int_{-\infty}^{\frac{r}{\sqrt{n}}} e^{-y^2/2} dy \tag{1.29}$$

If $n_1 < n_2 < n_3 < n_4$, then $(x_{n_2T} - x_{n_1T})$ is independent of $(x_{n_4T} - x_{n_3T})$; that is, the number of heads from the n_1th to n_2th toss is independent of the number of heads from the n_3th to n_4th toss.

Letting $t = nT$,

$$E\{x_t\} = 0 \tag{1.30}$$

$$E\{x_t^2\} = \frac{tS^2}{T}. \tag{1.31}$$

Suppose t is fixed and let $S \to 0$ and $T \to 0$. That is, the step size becomes smaller and smaller as the time between tosses approaches zero. The variance of x_t will remain constant only if S tends to zero as $\sqrt{T}$. Therefore, let

$$S^2 = \sigma^2 T \tag{1.32}$$

and write the variance as

$$E\{x_t^2\} = \sigma^2 t \tag{1.33}$$

Defining

$$w_t = \lim_{T \to 0} x_t, \tag{1.34}$$

then w_t is the Wiener stochastic process.

The Wiener process has the following properties[5]:

1. $w_0 = 0$, and w_t is real.
2. w_t is Gaussian with zero mean and variance $\sigma^2 t$.
3. The process has independent increments.
4. The process has stationary increments – that is, the density of $\left(w_{t_1} - w_{t_2}\right)$ depends only on $(t_1\text{-}t_2)$ – but w_t itself is not stationary.
5. The variance of $\left(w_{t_1} - w_{t_2}\right)$ is proportional to $(t_1 - t_2)$:

   $$E\left\{(w_{t_1} - w_{t_2})^2\right\} = \sigma^2(t_1 - t_2).$$

6. The process is not differentiable in the mean square sense.

STOCHASTIC DIFFERENTIATION

Define

$$x_t' = \frac{dx_t}{dt} = \lim_{\varepsilon \to 0} \frac{x_{t+\varepsilon} - x_t}{\varepsilon} \tag{1.35}$$

If the above limit exists, then x_t' has the usual meaning of a derivative. The process x_t has a mean square derivative at t if we can find another process x_t' with the property that

$$\lim_{\varepsilon \to 0} E\left\{\left[\frac{x_{t+\varepsilon} - x_t}{\varepsilon} - x_t'\right]^2\right\} = 0 \tag{1.36}$$

This criterion may be difficult to apply in practice because the process x_t' is not, in general, given, and must be guessed at. An alternative criterion (Cauchy) shows that the mean square derivative exists if

$$\lim_{\varepsilon_1,\, \varepsilon_2 \to 0} E\left\{\left[\frac{x_{t+\varepsilon_1} - x_t}{\varepsilon_1} - \frac{x_{t-\varepsilon_2} - x_t}{\varepsilon_2}\right]^2\right\} = 0. \tag{1.37}$$

It can be shown that a second order stochastic process x_t is differentiable in the mean square sense if[6]

$$\frac{\partial^2 r(t_1, t_2)}{\partial t_1 \partial t_2} \tag{1.38}$$

exists at $t_1 = t_2$, and where $r(t_1, t_2)$ is the *autocovariance function* of the process. That is

$$r(t_1, t_2) = E\left\{\left[x_{t_1} - \mu_{t_1}\right]\left[x_{t_2} - \mu_{t_2}\right]\right\} \tag{1.39}$$

For the Wiener process with $t_2 > t_1$

$$\begin{aligned} E\left\{\left[w_{t_2} - w_{t_1}\right]\left[w_{t_1} - w_0\right]\right\} &= E\left\{\left[w_{t_2} - w_{t_1}\right]w_{t_1}\right\} \\ &= E\left\{w_{t_2} - w_{t_1}\right\}E\left\{w_{t_1}\right\} = 0 \end{aligned} \tag{1.40}$$

because $w_0 = 0$ and using the independent increments property. Therefore,

$$E\left\{\left[w_{t_2} - w_{t_1}\right]w_{t_1}\right\} = E\left\{w_{t_2} w_{t_1}\right\} - E\left\{w_{t_1}^2\right\} = 0 \tag{1.41}$$

and

$$E\left\{w_{t_1} w_{t_2}\right\} = E\left\{w_{t_1}^2\right\} = \sigma^2 t_1 \tag{1.42}$$

The autocovariance function can now be written as

$$r(t_1, t_2) = \begin{cases} \sigma^2 t_2 & \text{for } t_1 \geq t_2 \\ \sigma^2 t_1 & \text{for } t_2 \geq t_1 \end{cases} \tag{1.43}$$

Next we take the partial derivatives of $r(t_1, t_2)$ in order to determine if the Wiener process is differentiable in the mean square sense.

$$\frac{\partial r(t_1, t_2)}{\partial t_1} = \begin{cases} 0 & t_1 \geq t_2 \\ \sigma & t_2 \geq t_1 \end{cases} \tag{1.44}$$

and therefore,

$$\frac{\partial^2 r(t_1, t_2)}{\partial t_1 \partial t_2} \text{ does not exist.}$$

We conclude that although the Wiener process has continuous sample paths in the mean square sense, it is not anywhere differentiable in the mean square sense.

PURE WHITE NOISE

Time series statisticians make frequent use of a process called *pure white noise*. This process has independent increments and is therefore serially uncorrelated. In continuous time, pure white noise is represented mathematically as the time derivative of a Wiener process.

$$n_t = w_t' \tag{1.45}$$

where n_t is the noise. We know from the previous section that the Wiener process is not differentiable in the mean square sense. Nevertheless, it is convenient to regard n_t as a mathematical representation of a process with infinite variance and independent increments.

In discrete time, however, pure white noise exists as a well-defined stochastic process. We will make frequent use in the text of discrete time pure white noise and will usually represent it symbolically as ε_t or η_t. Discrete time pure white noise is said to be "delta correlated." That is

$$E\{\eta_i \eta_j\} = r(t_i, t_j) = \Delta_{ij} \tag{1.46}$$

where

$$\Delta_{ij} = \begin{cases} 1, & i = j \\ 0, & i \neq j \end{cases} \tag{1.47}$$

is the Kronecker delta function.

Continuous time pure white noise does not have finite variance, and therefore, the definition of its autocovariance function is a bit problematic. We formally write the autocovariance function as

$$E\{n_t n_\tau\} = r(t, \tau) = \delta(\tau - t) \tag{1.48}$$

where

$$\delta(\tau - t) = \begin{cases} \infty, & \tau = t \\ 0, & \tau \neq t \end{cases} \tag{1.49}$$

and

$$\int_{-\infty}^{\infty} \delta(s)ds = 1 \tag{1.50}$$

is the Dirac delta function. Therefore, both discrete and continuous time pure white noise are delta correlated.

It is important to distinguish between pure white noise and just *white noise.* The white noise process is delta correlated, as is pure white noise. However, white noise is not an independent increments process. That is to say, white noise can be serially dependent even though it is uncorrelated. An example of this distinction is provided in a later section of the chapter.

STOCHASTIC INTEGRATION

Consider the Riemann-Stieltjes time integral

$$S = \int_a^b x_t dt \tag{1.51}$$

If this integral exists for every realization of x_t, it defines a random number, S, and the integral has the usual meaning of the area under x_t. Stochastic integrals can be defined in the mean square sense as a limit of the expectation of the square of a Riemann sum:

$$\lim_{\Delta t_i \to 0} E\left\{ |S - \sum_i x_{t_i}(\omega)\Delta t_i|^2 \right\} \tag{1.52}$$

If the limit is zero then S is defined at the integral of x_t in the mean square sense.

Unfortunately, Riemann-Stieltjes stochastic integrals of the form

$$\int f(\omega_t, t)d\omega_t \tag{1.53}$$

where $\{w_t, t \in T\}$ is a Wiener process do not exist in the mean square Riemann-Stieltjes sense because, although the process w_t is continuous with probability one, almost all of its sample functions have unbounded variation.

ÎTO PROCESSES

To better understand why the integral in (1.53) cannot be interpreted as an ordinary Riemann-Stieltjes intragral, consider a stochastic process x_t purportedly defined by the stochastic differential equation

$$\frac{dx_t}{dt} = f(x_t, t) + g(x_t, t)n_t \qquad t \geq t_0 \qquad (1.54)$$

with initial condition x_{t_0} which is independent of n_t, where n_t is a pure white noise process, and where $g(x_t, t)$ is some nonzero function. Attempting to integrate both sides of (1.54) one is stymied by the fact that the sample functions of the process n_t are delta functions, so that $g(x_t, t)n_t$ is not mean square Riemann integrable.

However, recall from above that n_t is the formal derivative of a Wiener process:

$$n_t \sim \frac{dw_t}{dt} \qquad (1.55)$$

so the stochastic differential equation of the previous section can be written

$$dx_t = f(x_t, t)dt + g(x_t, t)dw_t \qquad t \geq t_0 \qquad (1.56)$$

and is meaningful if and only if its integral is defined:

$$x_t - x_{t_0} = \int_{t_0}^{t} f(x_\tau, \tau)d\tau + \int_{t_0}^{t} g(x_\tau, \tau)dw_\tau \qquad (1.57)$$

The first integral can be defined as a mean square Riemann-Stieltjes integral. The second integral has been defined in the mean square sense by Îto. Hence, (1.56) is called an "Îto stochastic differential equation." Its solution, if it exists, is x_t.

ILLUSTRATIVE EXAMPLES OF ÎTO PROCESSES

Our introduction to Îto processes begins with an example drawn from material in Section 2 in Chapter 4 of Jazwinski (1970).

Example (1.1): Consider

$$x_t = e^{w_t} \qquad t \geq 0 \tag{1.58}$$

and $w_0 = 0$ with probability 1. The process $\{w_t, t \geq 0\}$ is Wiener with variance parameter σ^2. Because w_t is not differentiable we can not differentiate (1.58) to obtain a differential equation. Instead, we write a power series expansion of (1.58):

$$\Delta x_t = e^{w_t + \Delta w_t} - e^{w_t} = e^{w_t}\left(e^{\Delta w_t} - 1\right) = x_t\left(e^{\Delta w_t} - 1\right)$$

$$e^{\Delta w_t} = 1 + \Delta w_t + \frac{(\Delta w_t)^2}{2} + \frac{(\Delta w_t)^3}{3!} + \ldots \tag{1.59}$$

Therefore,

$$\Delta x_t = x_t\left(\Delta w_t + \frac{(\Delta w_t)^2}{2} + \ldots\right) \tag{1.60}$$

If we truncate the expansion at the first-order term we have[7]:

$$E\{\Delta x_t - x_t \Delta w_t\} = E\left\{x_t\left(\frac{(\Delta w_t)^2}{2} + \frac{(\Delta w_t)^3}{3!} + \ldots\right)\right\}$$

$$= E\left\{E\left\{x_t\left(\frac{(\Delta w_t)^2}{2} + \ldots\right)|x_t\right\}\right\}$$

$$= E\left\{x_t E\left\{\left(\frac{(\Delta w_t)^2}{2} + \ldots\right)|x_t\right\}\right\}$$

$$= E\left\{x_t E\left\{\frac{(\Delta w_t)^2}{2} + \ldots\right\}\right\}$$

$$= E\{x_t\} E\left\{\frac{(\Delta w_t)^2}{2} + \ldots\right\}$$

$$= O(\Delta t)$$

and therefore, (1.61)

$$E\left\{|\Delta x_t - x_t \Delta w_t|^2\right\} = O(\Delta t)^2. \tag{1.62}$$

This follows because w_t is normal and hence

$$E\left\{(\Delta w_t)^n\right\} = \begin{cases} 0 & \text{all odd } n \geq 1 \\ 1 \cdot 3 \cdot 5 \ldots (n-1)(\sigma\sqrt{\Delta t})^n & \text{even } n \geq 2 \end{cases} \tag{1.63}$$

We conclude from the example that the representation of (1.58) by the differential equation

$$dx_t = x_t dw_t \tag{1.64}$$

involves errors of order dt and is consequently meaningless. Apparently the Δw_t^2 term in the expansion cannot be omitted.

Calculate[8]

$$E\left\{\Delta x_t - x_t \Delta w_t - \frac{x_t}{2}(\Delta w_t)^2\right\} = o(\Delta t) \tag{1.65}$$

so that

$$E\left\{|\Delta x_t - x_t \Delta w_t - \frac{x_t}{2}(\Delta w_t)^2|^2\right\} = o(\Delta t) \tag{1.66}$$

In the mean square sense we have the differential equation representation

$$dx_t = x_t dw_t + \frac{1}{2} x_t dw_t^2 \tag{1.67}$$

with $x_{t_0} = 1$.

Example (1.2): Compute the solution to (1.67). Multiplying (1.67) by the integrating factor e^{-w_t},

$$e^{-w_t} dx_t = x_t e^{-w_t} dw_t + \frac{1}{2} x_t e^{-w_t} dw_t^2 \tag{1.68}$$

Find $d(x_t e^{-w_t})$ using a power series expansion:

$$\begin{aligned} d(x_t e^{-w_t}) &= e^{-w_t} dx_t + x_t d(e^{-w_t}) \\ d(e^{-w_t}) &= e^{-w_t - dw_t} - e^{-w_t} = e^{-w_t}(e^{-dw_t} - 1) \end{aligned} \tag{1.69}$$

and

$$\begin{aligned} e^{-dw_t} - 1 &\cong 1 - dw_t - \frac{1}{2} dw_t^2 - 1 \\ &= -dw_t - \frac{1}{2} dw_t^2 \end{aligned} \tag{1.70}$$

Therefore

$$d(e^{-w_t}) = -e^{-w_t}\left\{dw_t + \frac{1}{2}dw_t^2\right\} \tag{1.71}$$

and

$$\begin{aligned} d(x_t e^{-w_t}) &= e^{-w_t}dx_t - x_t e^{-w_t}\left\{dw_t + \frac{1}{2}dw_t^2\right\} \\ &= e^{-w_t}dx_t - x_t e^{-w_t}dw_t - \frac{1}{2}x_x e^{-w_t}dw_t^2 \end{aligned} \tag{1.72}$$

Combining this with (1.68) we have

$$d\left(x_t e^{-w_t}\right) = 0 \tag{1.73}$$

so that

$$x_t = ce^{w_t} \tag{1.74}$$

In review of the initial condition it must be that c = 1, and we get

$$x_t = e^{w_t} \tag{1.75}$$

as the solution to (1.67), thus concluding the example.

Îto's Lemma is a rule that specifies how to take the differential of a function of an Îto process. Below we state it without proof for the case where x_t is a scalar process:

Îto's Lemma: Let x_t be the unique solution to the stochastic differential equation

$$dx_t = f(x_t, t) + g(x_t, t)dw_t \qquad t \geq t_0 \tag{1.76}$$

where w_t is a Wiener process with $E\{dw_t^2\} = dt$. Let $\Phi(x_t, t)$ be a real function, continuously differentiable in t and having continuous second mixed partial derivatives with respect to x. Then the stochastic differential $d\Phi$ of Φ is

$$d\Phi = \Phi_t dt + \Phi_x dx_t + \frac{1}{2}g^2\Phi_{xx}dt \tag{1.77}$$

where subscripts on Φ denote partial derivatives.

Modern-day finance theory often assumes that stock market prices follow a geometric random walk with drift in continuous time. If P_t is the stock price at time t, then the differential of P_t is written

$$dP_t = \alpha P_t dt + \sigma P_t dw_t \tag{1.78}$$

The next example solve s(1.78) for P_t.

Example (1.3): Show that the solution to (1.78) is given by

$$P_t = P_{t_0} e^{(\alpha - \sigma^2/2)(t - t_0) + \sigma(w_t - w_{t_0})} \tag{1.79}$$

Note: because dP_t is normal, P_t is log-normal in distribution.

In terms of Itô's Lemma, set

$$dx_t = dw_t \tag{1.80}$$

which implies $x_t = w_t$, and

$$f(x_t, t) = 0 \tag{1.81}$$

$$g(x_t, t) = 1 \tag{1.82}$$

Writing P_t as

$$P_t = \Phi(x_t, t) = \Phi(w_t, t) = P_{t_0} e^{(\alpha - \sigma^2/2)(t - t_0) + \sigma(w_t - w_{t_0})} \tag{1.83}$$

We now have

$$\Phi_t = (\alpha - \sigma^2/2)P_t = (\alpha - \sigma^2/2)P_{t_0} e^{(\alpha - \sigma^2/2)(t - t_0)} \tag{1.84}$$

$$\Phi_x = \Phi_w = \sigma P_t = \sigma P_{t_0} e^{(\alpha - \sigma^2/2)(t - t_0) + \sigma(w_t - w_{t_0})} \tag{1.85}$$

$$\Phi_{xx} = \Phi_{ww} = \sigma^2 P_t = \sigma^2 P_{t_0} e^{(\alpha - \sigma^2/2)(t - t_0) + \sigma(w_t - w_{t_0})} \tag{1.86}$$

Putting everything together and collecting terms yields

$$d\Phi = dP_t = (\alpha - \sigma^2/2)P_t dt + \sigma P_t dw_t + \frac{1}{2}(1)^2\sigma^2 P_t dt$$
$$= \alpha P_t dt + \sigma P_t dwt, \tag{1.87}$$

completing the example.

Sometimes equation (1.78) is written as:

$$\frac{dP_t}{P_t} = \alpha dt + \sigma dw_t \tag{1.88}$$

In this form (dP_t/P_t) is regarded as the instantaneous rate of return with

$$E\left\{\frac{dP_t}{P_t}\right\} = \alpha dt \tag{1.89}$$

$$Var\left\{\frac{dP_t}{P_t}\right\} = \sigma^2 dt \tag{1.90}$$

One might be tempted to solve equation (1.88) by integration:

$$\int_{t_0}^{t} d\frac{P_\tau}{P_\tau} = \int_{t_0}^{t} \alpha d\tau + \int_{t_0}^{t} \sigma dw_\tau \tag{1.91}$$

so that

$$\ln P_t - \ln P_{t_0} = \alpha(t - t_0) + \sigma(w_t - w_{t_0}) \tag{1.92}$$

and

$$\frac{P_t}{P_{t_0}} = e^{\alpha(t - t_0) + \sigma(w_t - w_{t_0})}. \tag{1.93}$$

This is clearly incorrect because if we apply Itô's Lemma to equation (1.93) the result is not equation (1.78). The reason why this is the case should be fairly obvious. The stochastic differential equation

$$dP_t = \alpha P_t dt + \sigma P_t dw_t \tag{1.94}$$

is a shorthand way of writing

$$P_t - P_{t_0} = \int_{t_0}^{t} \alpha P_\tau d\tau + \int_{t_0}^{t} \sigma P_\tau dw_\tau \tag{1.95}$$

The second integral is not defined in the mean square Riemann sense because P_t is a function of ω_t, although it can be defined in the Itô calculus. However if

$$dx_t = x_t dt + \sigma dw_t \tag{1.96}$$

then the formalistic rules of ordinary calculus will work.

$$\frac{dx_t}{dt} = x_t + \sigma \frac{dw_t}{dt} \tag{1.97}$$

$$dx_t - x_t dt = \sigma dw_t$$

By integration the solution is

$$x_t = \sigma \int_{t_0}^{t} e^{(t-s)} dw_s + x_t e^{t} \tag{1.98}$$

NONLINEAR STOCHASTIC PROCESSES

Suppose a time series is generated by a continuous-time *linear* model,

$$x_t = \int_{-\infty}^{\infty} h(\alpha)\, n_t(t - \alpha)\, d\alpha \tag{1.99}$$

where $\{n_t, t \in T\}$ is the pure white noise process developed earlier. The above model (1.99) is a fairly general representation of stochastic processes having a continuous spectra and a defined probabilistic structure. In physically realizable linear systems, the weighting function h(t), called the *impulse response*, is real, and it equals zero for negative t (causality). If $\{n_t, t \in T\}$ is Gaussian, then $\{x_t, t \in T\}$ will be Gaussian. Alternatively, if $\{n_t\}$ is pure white noise, but non-Gaussian, then $\{x_t\}$ will be non-Gaussian. Note that (1.99) is not intended to be an exhaustive definition of the linear model since, for example, a wide variety of point processes are also linear.

The model in (1.99) will be *nonlinear* if the input process n_t alters the impulse response function h(t). For example, let n_L be a sequence of past values of n_t, up to and including t - L. Then we might write the impulse response function as

$$h(n_L, \alpha) = h(\alpha) + \int_0^L a(\beta, \alpha) n_{t-\beta-\alpha} d\beta \tag{1.100}$$

where $a(\beta, \alpha)$ is another weighting function that is convolved with the input process. If we replace $h(\alpha)$ in (1.99) with (1.100), x_t will be nonlinear even if n_t is Gaussian and independently distributed.

The vast majority of nonlinear time series models are expressed as discrete time models. The most widely studied nonlinear discrete time models are the Auto Regressive Conditionally Heteroskedastic or ARCH models (see Engle (1982)) and their close cousins, the Generalized Autoregressive Conditionally Heteroskedastic or GARCH models (see Bollerslev (1986)). These two classes of models have had wide appeal to time series statisticians because of their ability to model time varying volatility in a stochastic process.

All the ARCH/GARCH specifications have the following structure:

$$x_t = \sqrt{h_t}\, \eta_t \tag{1.101}$$

where h_t is the functional

$$h_t = h(\eta_{t-1}, \eta_{t-2}, \ldots), \tag{1.102}$$

$\{\eta_t\}$ is discrete time pure white noise, and h(.) is a positive function on $\{\eta_{t-1}, \eta_{t-2}, \ldots \eta_{t-s}\}$.

For example, a GARCH (1,1) model might be

$$x_t = \sqrt{h_t}\, \eta_t \tag{1.103}$$

where

$$h_t = \left\lfloor b_0 + b_1\eta_{t-1}^2 + b_3 h(t-1) \right\rfloor \tag{1.104}$$

with b_0, b_1, and b_3 positive parameters. Most GARCH models have high persistence in their volatility; that is to say, a volatility shock does not die out very quickly. Usually it is assumed that $\{\eta_t\}$ is Gaussian.

Next we briefly list some other examples of nonlinear time series models. Parameters will be denoted as c_1, c_2, ... (Often there are restrictions on the parameters to ensure stability, and where appropriate, invertibility.)

(a) *Bilinear Model*

$$x_t = \eta_t + c_1\eta_{t-1} x_{t-2} \tag{1.105}$$

(b) *Nonlinear Moving Average* (MA)

$$x_t = \eta_t + c_1\eta_{t-1}\eta_{t-2} \tag{1.106}$$

(c) *Extended Nonlinear Moving Average* (MA)

$$x_t = \eta_t + c_1\eta_{t-1} \sum_{j=2}^{L} (c_1)^{j-2}\eta_{t-j} \tag{1.107}$$

(d) *Nonlinear Autoregressive* (AR)

$$x_t = [c_1 + c_1\eta_{t-1}]x_{t-1} + \eta_t \tag{1.108}$$

(e) *Threshold* AR

$$\begin{aligned} x_t &= -c_1 x_{t-1} + \eta_t && \text{if } x_{t-1} \le 1 \\ x_t &= c_2 x_{t-1} + \eta_t && \text{otherwise} \end{aligned} \tag{1.109}$$

(f) *Nonlinear Threshold* AR

$$x_t = -(c_1 + c_2|x_{t-1}|)x_{t-1} + \eta_t \qquad \text{if } |x_{t-1}| \leq 1$$
$$x_t = -.5x_{t-1} + \eta_t \qquad \text{otherwise} \tag{1.110}$$

(g) *Exponential* AR

$$x_t = c_1 + c_2 \exp\left[-x_{t-1}^2\right]x_{t-1} + \eta_t \tag{1.111}$$

MARTINGALE DIFFERENCES

An important distinction exists between those stochastic processes which are *martingale differences*, and those which are not. Let $\{x_t\}$ be a real-valued process with discrete or continuous parameter set. The process is a martingale difference if $E\{|x_t|\} < \infty$ for all t, and if for any $t > t_1$, $E\{x_t|x_{t_1} = a\} = 0$ for any value of a. It follows from the law of conditional expectations that the unconditional expectation of x_t is zero.

Furthermore x_t is uncorrelated with any function of the past observations, say $f(x_{t-1})$. This is easy to see because

$$E\{x_t f(x_{t-1})|x_{t-1}\} = f(x_{t-1})E\{x_t|x_{t-1}\} = 0 \tag{1.112}$$

and therefore the unconditional expectation of $x_t f(x_{t-1})$ is also zero.

Martingale differences are related to the concept of *white noise* and *pure white noise*. White noise is serially uncorrelated, whereas pure white noise satisfies a stronger assumption: it is both identically and independently distributed. The relationship between a martingale difference and white noise can be summarized in the following two statements:

(1) All martingale differences are uncorrelated, but not all white noise processes are martingale differences.

(2) All zero mean independent processes are martingale differences, but the converse does not follow.

The essential feature of a martingale difference process is that it does not admit forecasts of the conditional mean; that is, one can not make point forecasts using past observations. We will have more to say about forecasting in the next section of this chapter. The ARCH/GARCH models are examples of martingale differences, whereas the bilinear model is not. However, note that both ARCH/GARCH and the bilinear models are examples of white noise – they are serially uncorrelated.

Figure 1-2 is a plot of an independent, zero mean Gaussian process, η_t, with unit variance; it is also a martingale difference. That is, $\eta_t \sim \mathrm{NIID}(0, 1)$. Next this process is used as the innovation series for an extended nonlinear moving average model:

$$\begin{aligned} x_t = \eta_t &+ (0.5)(-0.429\eta_{t-1}\eta_{t-17} + 0.57\eta_{t-2}\eta_{t-25} \\ &+ 0.105\eta_{t-3}\eta_{t-27} + 0.509\eta_{t-4}\eta_{t-29} - 0.657n_{t-5}\eta_{t-2} \\ &+ 0.683\eta_{t-6}\eta_{t-9} - 0.949\eta_{t-7}\eta_{t-23} + 0.703\eta_{t-8}\eta_{t-21} \\ &- 0.694\eta_{t-9}\eta_{t-14} + 0.872\eta_{t-11}\eta_{t-17} + 0.489\eta_{t-14}\eta_{t-25} \\ &+ 0.902\eta_{t-16}\eta_{t-27} - 0.954\eta_{t-20}\eta_{t-28} - 0.992\eta_{t-24}\eta_{t-32} \\ &- 0.349\eta_{t-26}\eta_{t-41}) \end{aligned} \tag{1.113}$$

A plot of $\{x_t\}$ is shown in Figure 1-3. It is not a martingale difference process. In both plots we have added horizontal dashed lines to indicate the limits of a 95% confidence interval – i.e., ± two estimated standard deviations – around the observations. The relatively large number of observations falling outside these limits is indicative of the non-Gaussian nature of this process. The distributions of data generated by nonlinear processes are often characterized by "fat tails" compared to those of a Gaussian distribution. We illustrate this characteristic by providing histograms of $\{\eta_t\}$ and $\{x_t\}$ in Figures 1-4 and 1-5.

Figure 1-2 Realization of a Pure White Noise Process
with Unit Variance

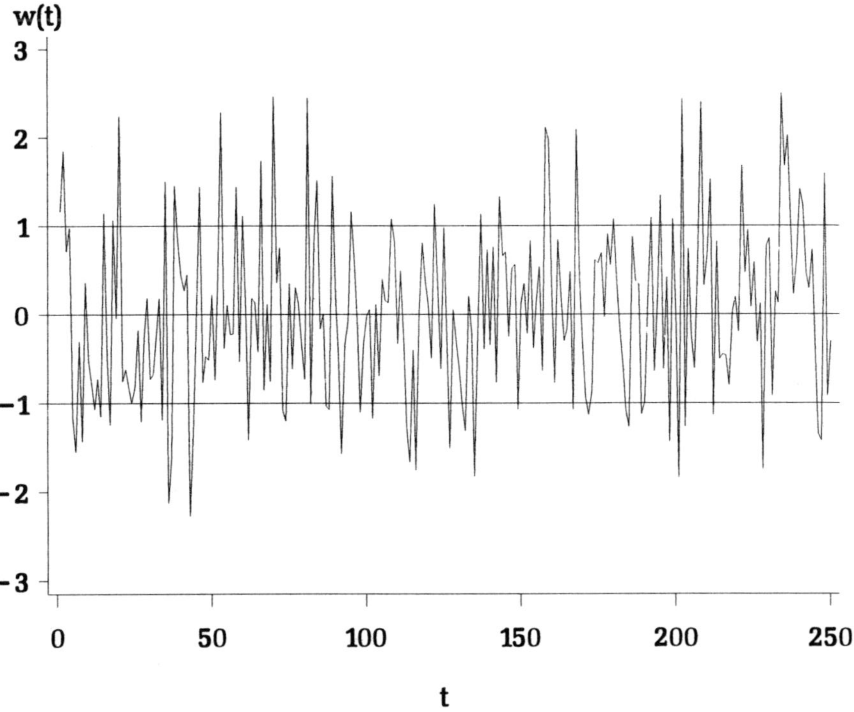

WHY DOES THE PRESENCE OF NONLINEARITY IN TIME SERIES DATA MATTER?

There are three good reasons why the presence of nonlinearity in time series data may cause problems in data analysis:

(1) Certain estimates may be poorly behaved.

(2) A linear model fitted to the data may leave out important dependencies.

(3) A nonlinear process that is also white noise may be forecastable.

As an example of (1) consider the estimation of the unconditional variance of a time series. The estimator is derived under the assumption that the data are independent normal random variables with common mean and variance. Time series data often violate the independence assumption. In the case of autocorrelation the correction to the estimator is straightforward: after a suitable linear filter is applied to correlated data one is left with a white noise residual series. But what if these white – i.e., uncorrelated – residuals are nevertheless still serially dependent because the generating process was nonlinear?

We illustrate what can happen with a simple example. Two time series of length 2000 were generated using a Gaussian pseudo random number generator. The first series, $\{\eta_t\}$, is Gaussian pure white noise with unit variance. The second series, $\{x_t\}$, is a nonlinear moving average with 15 lag terms generated from equation 1.113 above. The $\{x_t\}$ series was Studentized, thereby forcing its sample variance to unity. Next, each series was divided into subgroups of 100 observations each and the sample variance of each subgroup was calculated. In Table 1-1, we list the sample variances for each subgroup. For the pure white noise series, $\{\eta_t\}$, an estimated 95% confidence interval is also given. Note the poor sample behavior of the variance estimates for the nonlinear $\{x_t\}$ series; these estimates vary much more widely about the correct value (essentially 1.0) than do the variance estimates for the $\{\eta_t\}$ series. Furthermore, note that more than half of the $\{x_t\}$ variance estimates fall outside the 95% confidence interval.

Figure 1-3 Realization of a 15-Term Quadratic Moving Average Process with Unit Variance

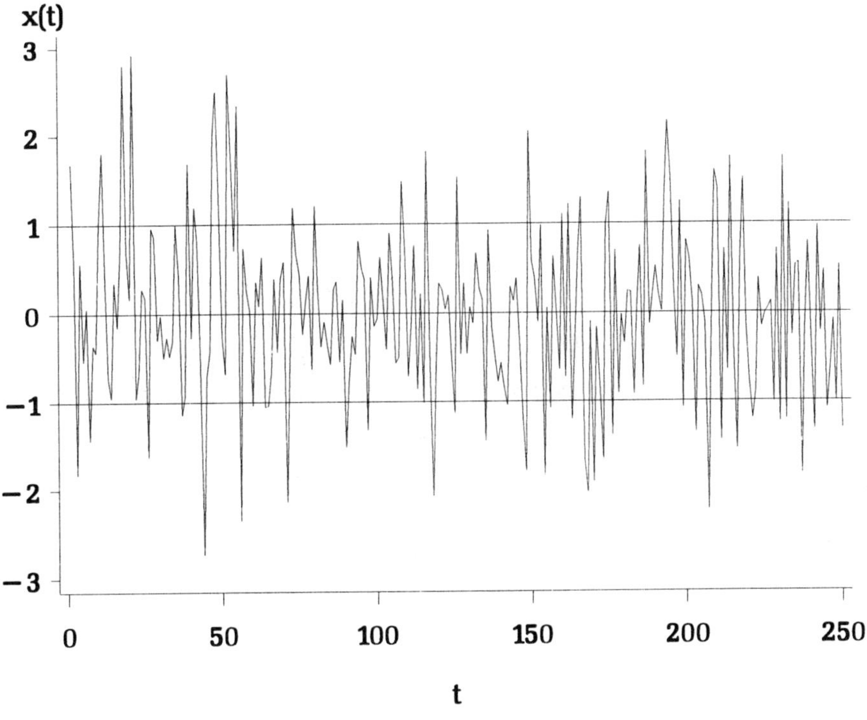

Figure 1-4 Histogram of a Pure White Noise Process with Unit Variance

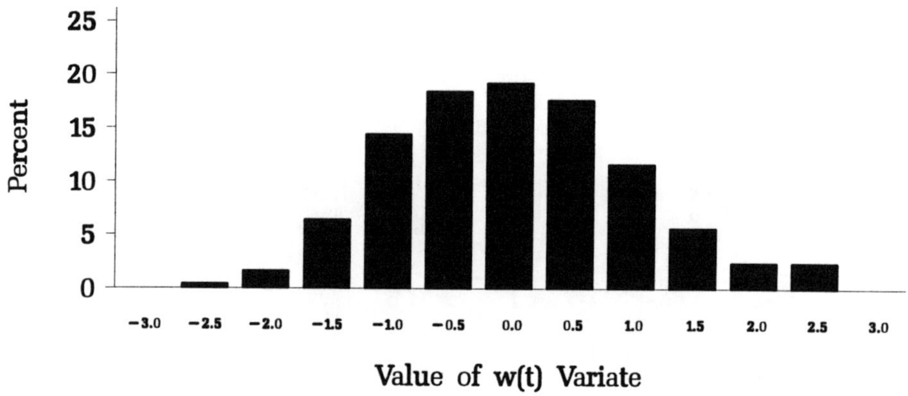

Figure 1-5 Histogram of a 15-Term Nonlinear Quadratic Moving Average Process With Unit Variance

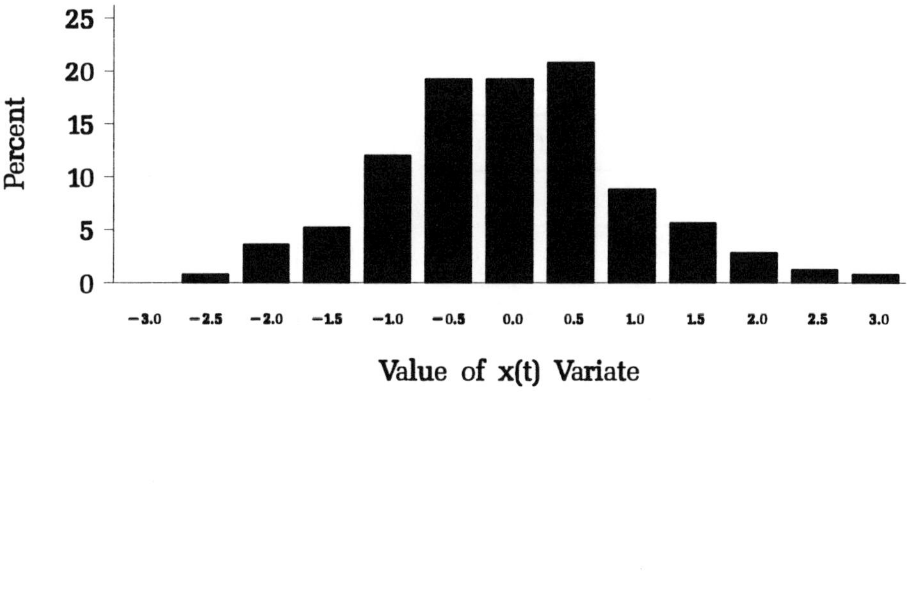

Table 1-1 Comparison of the Sample variances from a Pure White Noise Process and a Nonlinear White Noise Process[a]

Subgroup	Pure White Noise	Nonlinear White Noise
1	1.035 (0.835, 1.395)	0.563
2	0.714 (0.576, 0.962)	0.500
3	1.077 (0.869, 1.451)	2.461
4	1.235 (0.996, 1.664)	1.927
5	0.997 (0.804, 1.345)	1.297
6	1.145 (0.923, 1.543)	1.204
7	0.947 (0.764, 1.276)	0.994
8	1.033 (0.833, 1.392)	0.810
9	1.069 (0.862, 1.44)	1.423
10	0.788 (0.635, 1.062)	0.483
11	0.832 (0.671, 1.121)	0.899
12	1.174 (0.947, 1.582)	1.626
13	0.974 (0.785, 1.313)	0.583
14	0.967 (0.780, 1.303)	0.994
15	0.938 (0.780, 1.264)	0.773
16	0.849 (0.685, 1.144)	0.396
17	1.096 (0.884, 1.477)	0.865
18	0.840 (0.677, 1.132)	0.697
19	0.989 (0.798, 1.333)	0.712
20	1.083 (0.874, 1.460)	0.683

[a]The population variance is essentially 1.0 for each series. The numbers in parentheses are the limits of a 95% confidence interval for the variance based on the 100 sample observations in each subgroup.

We next turn to the second reason why nonlinearity is important. Consider the following autoregressive model of order 1:

$$x_t = \varphi x_{t-1} + u_t \tag{1.114}$$

with $|\varphi| < 1$ and $\{u_t\}$ a stationary white noise series. Then, the conditional expectation of x_t

$$E\{x_t | x_s, \ s \leq t-1\} = \varphi x_{t-1} \tag{1.115}$$

if and only if

$$E\{u_t | u_s, \ s \leq t-1\} = 0, \tag{1.116}$$

that is, $\{u_t\}$ must be a martingale difference process. Now suppose instead that the error sequence $\{u_t\}$ is generated by the following invertible nonlinear moving average model:

$$u_t = \eta_t + \sum_{m=1}^{L} a(m)\eta_{t-1}\eta_{t-m-1} \tag{1.117}$$

where $\{\eta_t\}$ is a pure white noise process – independent and identically distributed (i.i.d.) with zero mean. For this particular model

$$E\{u_t | u_s, \ s \leq t-1\} = \sum_{m=1}^{L} a(m)\eta_{t-1}\eta_{t-m-1} \tag{1.118}$$

and therefore,

$$E\{x_t | x_s, \ s \leq t-1\} = \Phi x_{t-1} + \sum_{m-1}^{L} a(m)\eta_{t-1}\eta_{t-m-1} \tag{1.119}$$

which contradicts equation (1.114).[9]

The above example leads to the third reason as to why nonlinearity is an important aspect of time series analysis. The ability of a nonlinear white noise process to produce forecasts can be illustrated with reference to the bilinear model:

$$x_t = b\,\eta_{t-1}x_{t-2} + \eta_t \tag{1.120}$$

where $\{\eta_t\}$ is again pure white noise. {This bilinear model is discussed in Granger and Andersen (1978), Priestley (1981), and Ashley and Patterson (1986).} This model is invertible if $b^2\sigma_n^2 < ½$. Because η_{t-1} and x_{t-2} are known at time t, it follows directly that the conditional mean of x_t is

$$E\{x_t | x_{t-1}, x_{x-2}, \ldots\} = b\eta_{t-1}x_{t-2}. \tag{1.121}$$

Yet, x_t is a white noise process so it is not forecastable using linear models. First we show that the unconditional mean of x_t is zero. Note that η_{t-1} is independent of x_{t-2} because it is a *future* i.i.d. innovation. Then

$$\begin{aligned} E\{x_t\} &= E\{b\eta_{t-1}x_{t-2} + \eta_t\} \\ &= bE\{\eta_{t-1}\}\,E\{x_{t-2}\} + E\{\eta_t\} \\ &= 0 \end{aligned} \tag{1.122}$$

Thus, the autocovariance of x_t at lag k is

$$\operatorname{cov}(x_t, x_{t-k}) = E\{x_t x_{t-k}\} \tag{1.123}$$

If $k > 1$, then x_{t-k} is not a function of η_t. Consequently,

$$\begin{aligned} \operatorname{cov}(x_t, x_{t-k}) &= E\{(b\eta_{t-1}x_{t-2} + \eta_t)x_{t-k}\} \\ &= E\{b\eta_{t-1}x_{t-2}x_{t-k}\} + E\{\eta_t x_{t-k}\} \\ &= bE\{\eta_{t-1}\}\,E\{x_{t-2}x_{t-k}\} + 0 \\ &= 0 \end{aligned} \tag{1.124}$$

Whereas, for k = 1,

$$
\begin{aligned}
E\{b\eta_{t-1}x_{t-2}x_{t-k}\} &= bE\{\eta_{t-1}x_{t-2}x_{t-1}\} \\
&= bE\{\eta_{t-1}x_{t-2}(b\eta_{t-2}x_{t-3} + \eta_{t-1})\} \\
&= b^2E\{\eta_{t-1}\eta_{t-2}x_{t-2}x_{t-3}\} + bE\{\eta_{t-1}^2 x_{t-2}\} \\
&= b^2E\{\eta_{t-1}\}E\{\eta_{t-2}x_{t-2}x_{t-3}\} + bE\{\eta_{t-1}^2\}E\{x_{t-2}\} \\
&= 0 \quad + \quad 0
\end{aligned}
\tag{1.125}
$$

Therefore,

$$\mathrm{cov}(x_t,\ x_{t-k}) = 0, \tag{1.126}$$

for $k \geq 1$, thus showing that the bilinear model is a white noise process that is nevertheless forecastable.

How good might the forecast be? One way of addressing this is through R^2, the proportion of the variance of the dependent variable, x_t, that can be "explained" by lagged values. For the problem at hand

$$R^2 = 1 - \frac{\mathrm{var}(\eta_t)}{\mathrm{var}(x_t)} \tag{1.127}$$

For the bilinear model

$$\mathrm{var}(x_t) = \frac{\sigma_\eta^2}{1 - b^2\sigma_\eta^2} \tag{1.128}$$

and therefore,

$$R^2 = 1 - [1 - b^2\sigma_\eta^2] = b^2\sigma_\eta^2 \tag{1.129}$$

Granger and Andersen (1978) show the $b^2\sigma_\eta^2$ must be less than 0.50 in order for the model to be invertible. Therefore, R^2 can be as high as 0.50 for this process when its nonlinear serial dependence is exploited, despite the fact that (since it is serially uncorrelated) it is not forecastable at all using linear time series methods.

References

Åström, K. (1970) *Introduction to Stochastic Control Theory* Academic Press: New York.

Ashley, R., Patterson, D. and M. Hinich (1986) "A Diagnostic Test for Nonlinear Dependence in Time Series Fitting Errors," *Journal of Time Series Analysis* Vol. 7, No. 3, pp. 165-178.

Bollerslev, T. (1986) "Generalized Autoregressive Conditional Heteroscedasticity," *Journal of Economctrics* 31, pp. 307-327.

Engle, R. (1982) "Autoregressive Conditional Heteroscedasticity with Estimates of the Variance of UK Inflation," *Econometrica* 50, pp. 987-1007

Feller, W. (1968) *An Introduction to Probability Theory and Its Applications* Vol. I, 3rd ed. John Wiley & Sons, Inc.: New York.

Granger, C. and A. Andersen (1978) *An Introduction to Bilinear Time Series Models* Vandenhuer and Ruprecht: Gottingen.

Jazwinski, A. (1970) *Stochastic Process and Filtering Theory* Academic Press: New York.

Papoulis, A. (1965) Probability, *Random Variables, and Stockastic Processes* McGraw-Hill: New York.

Priestley, M. (1981) *Spectral Analysis and Time Series* Academic Press: New York.

Rudin, W. (1964) *Principles of Mathematical Analysis*, 3rd ed. McGraw-Hill: New York.

Notes

1. See Rudin (1964) for a complete presentation of Borel fields and measure spaces.

2. See Jazwinski (1981), p. 39.

3. Convergence concepts for random sequences are covered in Jazwinski (1981), p. 57.

4. See Feller (1968), Chapter 7, for a discussion of the normal approximation to the binomial distribution.

5. See Papoulis (1965), p. 503.

6. See Åström (1970), p. 37.

7. The notation O(Δt) means $\frac{O(\Delta t)}{\Delta t} \to M$ as $\Delta t \to 0$, $0 < M < \infty$.

8. By o(Δt) we mean that $\frac{o(\Delta t)}{\Delta t} \to 0$ as $\Delta t \to 0$.

9. Invertibility of the process given in equation 1.117 guarantees that η_t can be expressed in terms of x_t, x_{t-1}, etc.

2 DETECTING NONLINEAR SERIAL DEPENDENCE

INTRODUCTION

There are many statistical tests for nonlinear serial dependence in the literature. Some focus on a particular property characteristic of nonlinear processes, such as conditional heteroskedasticity; some focus on a particular parametric family of models, as in the way the Tsay test considers quadratic models. Others, such as the Hinich tests, focus on particular moments. Still others, such as the BDS test, focus on more general measures of relatedness.

Our intent here is to free the practitioner from the tyranny of the single test which happens to be readily available. Our aim is to provide a sufficient variety of tests as to both limit the risk of missing something important in the data and to create the possibility of obtaining useful information from the pattern of results across the various tests.[1] Accordingly, we have selected the following six tests for the Toolkit:

- Mcleod and Li (1983)
- Engle LM (1982)
- BDS (1996)
- Tsay (1986)
- Hinich and Patterson (1995) (bicovariance)
- Hinich (1982) (bispectrum)

Each of these tests is described in turn in the remaining sections of this Chapter.

Except for the Hinich bispectral test, these tests all share the same premise: once any *linear* serial dependence is removed from the data via a prewhitening model, any remaining serial dependence must be due to a nonlinear generating mechanism. Thus, each of these procedures is actually a test of serial independence applied to the (by construction) serially uncorrelated fitting errors of an AR(p) model for the sample data.[2] This fitting error series, standardized to zero mean and unit variance, is denoted by $\{x_t\}$ below.

In contrast, the Hinich bispectral test directly tests for a nonlinear generating mechanism, so prewhitening is not necessary in using this test. Moreover, Ashley, Patterson, and Hinich (1986) have shown that the test statistic for the Hinich bispectral test is invariant to linear filtering of the data, so – if prewhitening is done anyway – the adequacy of the prewhitening model is irrelevant to the validity of this test.

MCLEOD-LI TEST

This test for ARCH effects was proposed by McLeod and Li (1983) based on a suggestion in Granger and Andersen (1978). It looks at the autocorrelation function of the squares of the prewhitened data and tests whether corr(x_t^2 , x_{t-k}^2) is non-zero for some k. The autocorrelation function for the squared residuals $\{x_t^2\}$ is estimated by:

$$\hat{r}(k) = \frac{\sum_{=k+1}^{N} \left(x_t^2 - \hat{\sigma}^2\right)\left(x_{t-k}^2 - \hat{\sigma}^2\right)}{\sum_{t=1}^{T} \left(x_t^2 - \hat{\sigma}^2\right)}$$

where

$$\hat{\sigma}^2 = \sum_{t=1}^{N} \frac{x_t^2}{N}$$

Under the null hypothesis that x_t is an i.i.d process (and assuming that $E(x_t^8)$ exists) McLeod and Li (1983) show that, for fixed L:

$$\sqrt{N}\,\hat{r} = \left(\hat{r}(1), \ldots, \hat{r}(L)\right)$$

is asymptotically a multivariate unit normal. Consequently, for L sufficiently large, the usual Box-Ljung statistic

$$Q = N(N+2)\sum_{i=1}^{L} \frac{\hat{r}^2(k)}{N-i}$$

is asymptotically $\chi^2(L)$ under the null hypothesis of a linear generating mechanism for the data. Typically L is taken to be around 20; below results are quoted for L = 24, although the Toolkit computes the test for a variety of L values.

ENGLE LM TEST

This test was proposed by Engle (1982) to detect ARCH disturbances; as Bollerslev (1986) suggests, it should also have power against GARCH alternatives. As with most

LaGrange Multiplier tests, the test statistic itself is based on the R^2 of an auxiliary regression, in this case:

$$x_t^2 = \alpha_0 + \sum_{i=1}^{p} \alpha_k x_{t-i}^2 + v_t$$

Under the null hypothesis of a linear generating mechanism for x_t, NR^2 for this regression is asymptotically distributed $\chi^2(p)$.

BDS TEST

The BDS test is a nonparametric test for serial independence based on the correlation integral of the scalar series, $\{x_t\}$. For embedding dimension m, let $\{X_t^m\}$ denote the sequence of *m*-histories generated by $\{x_t\}$:

$$X_t^m = (x_t, \ldots, x_{t+m-1})'$$

where the prime denotes transpose.

Then the correlation integral $C_{m,T}(\varepsilon)$ for a realization of N is given by:

$$C_{m,N}(\varepsilon) = \sum_{t<s} I_\varepsilon(X_t^m, X_s^m) \left\{ \frac{2}{N_m (N_m - 1)} \right\}$$

where $N_m = N - (m - 1)$ and $I_\varepsilon(X_t^m, X_s^m)$ is an indicator function which equals one if the sup norm $\|X_t^m - X_s^m\| < \varepsilon$ and equals 0 otherwise. Basically, $C_{m,N}(\varepsilon)$ counts up the number of m-histories that lie within a hypercube of size ε of each other. Brock, Dechert, and Scheinkman (1996) exploit the asymptotic normality of $C_{m,N}(\varepsilon)$ under the null hypothesis that $\{x_t\}$ is an i.i.d. process to obtain a test statistic which asymptotically converges to a unit normal. This convergence requires extremely large samples for values of the embedding dimension (m) much larger than 2, so attention here is restricted to the cases m = 2, 3, and 4.[3] Where (as here) the data has been normalized to unit variance, the test is ordinarily computed for ε =.5, 1, and 2; here results are quoted for $\varepsilon = 1$.

TSAY TEST

The Tsay (1986) test is a generalization of the Keenan (1985) test; it explicitly looks for quadratic serial dependence in the data.

Let the $K = k(k-1)/2$ column vectors $V_1 \ldots V_K$ contain all of the possible crossproducts of the form $x_{t-i}\, x_{t-j}$, where $i \in [1, k]$ and $j \in [i, k]$. Thus, $v_{t,1} = x^2_{t-1}$; $v_{t,2} = x_{t-1}\, x_{t-2}$; $v_{t,3} = x_{t-1}\, x_{t-3}$; $v_{t,k+1} = x_{t-2}\, x_{t-3}$; $v_{t,k+2} = x_{t-2}\, x_{t-4}$, ... and $v_{t,K} = x^2_{t-k}$. And let $\hat{v}_{t,i}$ denote the projection of $v_{t,i}$ on the subspace orthogonal to $x_{t-1}, \ldots, x_{t-k}$ – i.e., the residuals from a regression of $v_{t,i}$ on $x_{t-1}, \ldots, x_{t-k}$.

The parameters $\gamma_1 \ldots \gamma_K$ are then estimated by applying OLS to the regression equation

$$x_t = \gamma_o + \sum_{i=1}^{K} \gamma_i \hat{v}_{ti} + \eta_t.$$

Note that the jth regressor in this equation is $\hat{v}_{t,j}$ – the period t fitting error from a regression of $v_{t,j}$ on $x_{t-1}, \ldots, x_{t-k}$. So long as p exceeds K, this projection is unnecessary for the dependent variable (x_t) since it is prewhitened using an AR(p) model[4]. The Tsay test statistic is then just the usual F statistic for testing the null hypothesis that $\gamma_1 \ldots \gamma_K$ are all zero.

HINICH BICOVARIANCE TEST

This test assumes that $\{x_t\}$ is a realization from a third-order stationary stochastic process and tests for serial independence using the sample bicovariances of the data. The (r,s) sample bicovariance is defined as:

$$C_3(r,s) = (N-s)^{-1} \sum_{t=1}^{N-s} x_t x_{t+r} x_{t+s} \qquad 0 \le r \le s.$$

The sample bicovariances are thus a generalization of a skewness parameter. The $C_3(r,s)$ are all zero for zero mean, serially i.i.d. data. One would expect non-zero values for the $C_3(r,s)$ from data in which x_t depends on lagged crossproducts, such as $x_{t-i}x_{t-j}$ and higher order terms.

Let $G(r,s) = (N-s)^{½}\, C_3(r,s)$ and define X_3 as

$$X_3 = \sum_{s=2}^{\ell} \sum_{r=1}^{s-1} [G(r,s)]^2.$$

Under the null hypothesis that $\{x_t\}$ is a serially i.i.d. process, Hinich and Patterson (1995) show that X_3 is asymptotically distributed $\chi^2(\ell\,[\ell-1]/2)$ for $\ell < N^{½}$; based on their simulations, they recommend using $\ell = N^{.4}$. The X_3 statistic detects non-zero third order correlations; it can be considered a generalization of the Box-Pierce portmanteau statistic.

HINICH BISPECTRAL TEST

This nonparametric test examines the third order moments (bicovariances) of the data in the frequency domain to obtain a direct test for a nonlinear generating mechanism, irrespective of any linear serial dependence which might be present. Results in Chapter 6 below indicate that this test is not terribly powerful. But when it does reject, one need not worry about the possibility that the linear prewhitening model has failed to remove all linear serial dependence in the data.

Suppose that $\{y_t\}$, the series of interest, is a third-order stationary time series with, for expositional convenience, $E[y_t] = 0$. The series $\{y_t\}$ might be serially correlated, in which case it is distinct from the prewhitened fitting error series denoted $\{x_t\}$ above. Letting $c_{yyy}(r,s)$ denote the third order cumulant function for $\{y_t\}$,

$$c_{yyy}(r,s) = E\left[y_t y_{t+r} y_{t+s}\right],$$

the bispectrum of $\{y_t\}$ at frequency pair (f_1, f_2) is its (double) Fourier transform:

$$B_y(f_1,f_2) = \sum_{r=-\infty}^{\infty} \sum_{s=-\infty}^{\infty} c_{yyy}(r,s) \exp[-i2\pi(f_1 r + f_2 s)].$$

$B_y(f_1, f_2)$ is a spatially periodic function of (f_1, f_2), whose principal domain is the triangular set $\Omega = \{0 < f_1 < ½,\ f_2 < f_1,\ 2f_1 + f_2 < 1\}$.[5]

The time series $\{y_t\}$ is linear if and only if it can be expressed as

$$y_t = \sum_{n=0}^{\infty} a(n)\, u(t-n),$$

where $u_t \sim iid(0, \sigma^2)$ and the weights $\{a(n)\}$ are fixed. The bispectrum of $\{y(t)\}$ has a particularly simple form when the generating mechanism for $\{y_t\}$ is linear:

Theorem 2-1: If $\{y_t\}$ is linear and the weights $\{a(n)\}$ are absolutely summable, then

1. $B_y(f_1, f_2) = \mu_3 A(f_1)A(f_2)A^*(f_1 + f_2)$ where

$$A(f) \equiv \sum_{s=0}^{\infty} a(s) \exp\left[-i2\pi fs\right],$$

$A^{*}(f)$ is its complex conjugate, and μ_3 is $E\{u(t)^3\}$.

and

2. the square of the skewness function, $\Psi^2(f_1, f_2)$,

$$\Psi^2(f_1,f_2) \equiv \frac{|B_y(f_1,f_2)|^2}{S_y(f_1)\, S_y(f_2)\, S_y(f_1+f_2)} = \frac{\mu_3^2}{\sigma^6}$$

is a constant for all frequency pairs (f_1, f_2) in Ω, where $S_y(f)$ is the spectrum of $\{y_t\}$ at frequency f.

This was first proven in Brillinger (1965); an elementary proof is given at the end of this chapter. Under the null hypothesis of a linear generating mechanism, Theorem 2-1 implies that the sample estimates of $\Psi^2(f_1, f_2)$ for different frequency pairs will differ from one another no more than one would expect due to sampling error. In particular, Hinich (1982) shows that consistent sample estimates of $2\Psi^2(f_1,f_2)$ are asymptotically distributed as a noncentral chi-squared variate, $\chi^2(2, \lambda)$, with constant non-centrality parameter (λ) under the null hypothesis of linearity; whereas, if the null hypothesis of linearity is false, then λ is dependent on f_1 and f_2. The Hinich test then uses an expression for the asymptotic distribution of the interdecile range of observations from a specified distribution given by David (1970) to test whether the dispersion in the estimates of $2\Psi^2(f_1,f_2)$ exceeds that which one would expect under the null hypothesis.[6]

The Hinich bispectral test has the nice property that it is unaffected by the application of a linear filter to y_t. This follows from the fact that the squared skewness function, $\Psi^2(f_1,f_2)$, is invariant to linear filtering:

Theorem 2-2 If $\{y_t\}$ satisfies the assumptions of Theorem 2-1 and $z_t = W(B)y_t$, where W(B) is a fixed, linear, causal filter with absolutely summable weights, then $\Psi_z^2(f_1, f_2)$ is identical to $\Psi_y^2(f_1, f_2)$.

Proof of Theorem 2-2:

This is proven in Ashley, Patterson, and Hinich (1986, p. 174), but it also follows directly from Theorem 2-1 since $A_z(f) = W(\exp\{i2\pi f\})A(f)$ so that the same factor of $W(\exp\{i2\pi f_1\})W(\exp\{i2\pi f_2\})W(\exp\{-i2\pi[f_1+f_2]\})$ appears in both the numerator and denominator of $\Psi_z^2(f_1, f_2)$ and hence cancels out.

Consequently – in contrast to all the other tests – the Hinich bispectral test is not confounded by linear serial dependence remaining in the data due to imperfect prewhitening.

Proof of Theorem 2-1:

Lemma 1: If $\{y_t\}$ is linear, then

$$B_y(f_1,f_2) = \mu_3 \sum_{n=-\infty}^{\infty} \sum_{m=-\infty}^{\infty} \exp\left[-i2\pi(f_1 m + f_2 n)\right] \sum_{j=n}^{\infty} a(j)a(j-n+m)a(j-n)$$

proof:

$$B_y(f_1,f_2) = \sum_{n=-\infty}^{\infty} \sum_{m=-\infty}^{\infty} E\{y_{t+n} y_{t+m} y_t\} \exp\left[-i2\pi(f_1 m + f_2 n)\right]$$

$$= \sum_{n=-\infty}^{\infty} \sum_{m=-\infty}^{\infty}$$

$$E\left\{\left[\sum_{j=0}^{\infty} a(j)u_{(t+n-j}\right]\left[\sum_{k=0}^{\infty} a(k)u_{t+m-k}\right]\left[\sum_{l=0}^{\infty} a(l)u_{t-l}\right]\right\} \exp\left[-i2\pi(f_1 m + f_2 n)\right]$$

$$= \sum_{n=-\infty}^{\infty} \sum_{m=-\infty}^{\infty} \exp\left[-i2\pi(f_1 m + f_2 n)\right]$$

$$\sum_{j=0}^{\infty} \sum_{k=0}^{\infty} \sum_{l=0}^{\infty} a(j)\, a(k) a(l)\; E\left\{u_{t+n-j} u_{t+m-k} u(t-l)\right\}$$

This expectation equals μ_3 if n-j = m-k = l; otherwise, it is zero. So l = j-n and k = j-n+m are the only values of l and k which lead to nonzero terms in the sums over l and k. Consequently, the expression for $B_y(f_1, f_2)$ becomes:

$$B_y(f_1,f_2) = \mu_3 \sum_{n=-\infty}^{\infty} \sum_{m=-\infty}^{\infty} \exp\left[-i2\pi(f_1 m + f_2 n)\right] \sum_{j=0}^{\infty} a(j)\, a(j-n+m) a(j-n)$$

but a(j-n) is zero for $j \in [0, n-1]$, so the sum over j can start at $j = n$. This completes the proof of Lemma 1.

Lemma 2: Assuming that the weights {a(n)} are absolutely summable, so that the sums defining A(f) converge,

$$A(f_1)A(f_2)\,A^*(f_1+f_2)$$
$$= \sum_{n=-\infty}^{\infty} \sum_{m=-\infty}^{\infty} \exp\left[-i2\pi(f_1 m + f_2 n)\right] \sum_{j=n}^{\infty} a(j)\, a(j-n+m) a(j-n)$$

$$A(f_1)A(f_2)\,A^*(f_1+f_2)$$

$$= \sum_{s=0}^{\infty} a(s)\exp\left[-i2\pi\, f_1 s\right] \sum_{\tau=0}^{\infty} a(\tau)\exp\left[-i2\pi\, f_2 \tau\right] \sum_{w=0}^{\infty} a(w)\exp\left[i2\pi\,(f_1+f_2) w\right]$$

$$A(f_1)A(f_2)\,A^*(f_1+f_2)$$

$$= \sum_{s=0}^{\infty} \sum_{\tau=0}^{\infty} \sum_{w=0}^{\infty} \exp\left[-i2\pi\left\{f_1[s-w] + f_2[\tau-w]\right\}\right] a(s)a(\tau)a(w)$$

Letting m = s-w, $m \in [-\infty, \infty]$ and letting n = τ-w, $n \in [-\infty, \infty]$,

$$A(f_1)A(f_2)A^*(f_1+f_2) = \sum_{m=-\infty}^{\infty} \sum_{n=-\infty}^{\infty} \sum_{w=0}^{\infty} \exp\left[-i2\pi\left\{f_1 m + f_2 n\right\}\right] a(m+w)a(n+w)a(w)$$

and letting j = n+w this becomes

$$A(f_1)A(f_2)A^*(f_1+f_2) = \sum_{m=-\infty}^{\infty} \sum_{n=-\infty}^{\infty} \sum_{j=n}^{\infty} \exp\left[-i2\pi\left\{f_1 m + f_2 n\right\}\right] a(m+j-n)\,a(j)\,a(j-n)$$

proving Lemma 2. Combining Lemma 1 and Lemma 2 proves part 1 of Theorem 2-1. Part 2 follows directly from the observation that

$$S_y(f) = \sigma^2 |A(f)|^2.$$

References

Ashley, R. and Patterson, D. M. (1989). "Linear Versus Nonlinear Macroeconomies" *International Economic Review 30*, 685-704.

Ashley, R., Patterson, D. M. and Hinich, M. (1986). "A Diagnostic Test for Nonlinear Serial Dependence in Time Series Fitting Errors" *Journal of Time Series Analysis 7*, 165-78.

Bollerslev, Tim (1986) "Generalized Autoregressive Conditional Heteroskedasticity" *Journal of Econometrics 31*, 307-27.

Brillinger, D. (1965) "An Introduction to Polyspectrum" *Annals of Mathematical Statistics 36*, 1351-74.

Brillinger, D. and M. Rosenblatt (1967) "Asymptotic Theory of kth Order Spectra" in*Spectral Analysis of Time Series*, (B. Harris, ed.) Wiley: New York, pp. 153-88.

Brock, W. A., Hsieh, D. A., and LeBaron, B.D. (1991) *A Test of Nonlinear Dynamics, Chaos,and Instability: Theory and Evidence* MIT Press: Cambridge.

Brock, W. A., Dechert W., and Scheinkman J. (1996) "A Test for Independence Based on the Correlation Dimension" *Econometric Reviews 15*, 197-235.

Dalle Molle, J. W. and Hinich M.J. (1995) "Trispectral Analysis of Stationary Random Time Series." *Journal of the Acoustical Society of America 97*, 2963-2978.

David, H. A. (1970) *Order Statistics* Wiley: New York.

Engle, Robert F. (1982) "Autoregressive Conditional Heteroskedasticity with Estimates of the Variance of United Kingdom Inflation" *Econometrica 50,* 987-1007.

Granger, C. W. J. and Andersen, A. A. (1978) *An Introduction to Bilinear Time Series Models* Vandenhoeck and Ruprecht: Gottingen.

Hinich, M. (1982) "Testing for Gaussianity and Linearity of a Stationary Time Series"*Journal of Time Series Analysis 3*, 169-76.

Hinich, M. (1996) "Testing for Dependence in the Input to a Linear Time Series Model" *Journal of Nonparametric Statistics 6*, 205-221.

Hinich, M. and Patterson D. M. (1985) "Evidence of Nonlinearity in Daily Stock Returns" *Journal of Business and Economic Statistics 3*, 69-77.

Hinich, M. and Patterson D. M. (1995) "Detecting Epochs of Transient Dependence in White Noise," unpublished manuscript.

Judge, G., W., Griffiths, C., Hill, H. L, Lütkepohl, Lee, T. C. (1985) *The Theory and Practice of Econometrics* John Wiley and Sons: New York.

Kaplan, D. T. (1993) "Exceptional Events as evidence for Determinism." *Physica D 73*, 38-48.

Keenan, D.M. (1985) "A Tukey Nonadditivity-type Test for Time Series Nonlinearity."*Biometrika 72*, 39-44.

McLeod, A. I. and Li, W. K. (1983) "Diagnostic Checking ARMA Time Series Models Using Squared-Residual Autocorrelations" *Journal of Time Series Analysis 4*, 269-73.

Mizrach, B. (1991) "A Simple Nonparametric Test for Independence." unpublished manuscript.

Nychka, D., Ellner, S., Gallant, A.R., and McCaffrey, D. (1992) "Finding Chaos in Noisy Systems."*Journal of the Royal Statistical Society B 54*, 399-426.

Ramsey, J.B. (1969) "Tests for Specification Errors in Classical Linear Least Squares Regression Analysis." *Journal of the Royal Statistical Society B 31*, 350-371.

Ramsey, J.B., C. L. Sayers and P. Rothman (1990) "The Statistical Properties of Dimension Calculations Using Small Data Sets: Some Economic Applications."*International Economic Review 31*, 991-1020.

Subba Rao, T. and Gabr, M. (1980) "A Test for Linearity of Stationary Time Series Analysis"*Journal of Time Series Analysis 1*, 145-58.

Tsay, Ruey S. (1986) "Nonlinearity Tests for Time Series" *Biometrika 73,* 461-6.

White, H. (1989) "Some Asymptotic Results for Learning in Single Hidden-Layer Feedforward Network Models." *Journal of the American Statistical Association 84*, 1003-1013.

Notes

1.It is obviously infeasible to include every known test in the Toolkit. Omitted tests include: Ramsey (1969), Ashley and Patterson (1986), Mizrach (1991), Nychka, et al. (1992), Kaplan (1993), and Dalle Molle and Hinich (1995). The most well known test not considered here is the neural net test of White (1989). This test is omitted because it has so many free parameters ("connection strengths") that White recommends choosing them at random. Since the test is then substantially different each time it is run, the best one can do is to run it several times and use the Bonferroni inequality to obtain an upper bound on the p-value for the test.

2.In our implementation p is chosen to minimize the Schwartz (SC) criterion. In contrast to alternative choices (e.g., AIC or FPE) the Schwartz criterion is known to be consistent for AR(p) order determination under the null hypothesis of a linear generating mechanism; see Judge, et al. (1985, p. 246).

3.See Ramsey, Sayers and Rothman (1990). The actual size of the test for N = 200 and 500 is examined in Chapter 4.

4.The value of K is user-selectable in the Toolkit; this parameter is set to 5 in the simulations reported below.

5.See Brillinger and Rosenblatt (1967) for a rigorous treatment of the bispectrum.

6.See Hinich (1982) and Ashley, Patterson, and Hinich (1986) for details. Based on simulation results in the latter paper, M is set to the integer closest to $N^{.6}$ in the calculations reported below. The interdecile (rather than the interquartile) range is used here because it yields test results which are more robust to non-Gaussianity in the data. See also Subba Rao and Gabr (1980) for an earlier approach.

3 HOW TO RUN THE TOOLKIT PROGRAM ON A PC

INTRODUCTION

A computer program which will carry out the tests described in Chapter 2 is available free of charge from the authors. It can be downloaded from either of two Internet locations:

http://ashleymac.econ.vt.edu/ashleyhome.html

http://www.wkap.nl/kapis/appendices/toolzipd.exe

The program package is in a self-extracting file named "toolzipd.exe." The package contains the: (1) executable Toolkit computer program, (2) the program's control file, and (3) a brief set of usage notes.

SYSTEM REQUIREMENTS

The following is required of the computer which is to run the program:

- A PC with a Pentium processor.
- Windows 95/98, or Windows NT.
- At least 6 Mbytes of free RAM.

GETTING STARTED

The following steps are necessary when running the program for the first time.

1. Uncompress the program package by executing the toolzip.exe file. This is can be done from Windows Explorer by double clicking on the file name, "toolzipd.exe."

2. Use a text editor or word processor program to open the toolkit.cnl file. Its contents should look like Figure 3-1. On line 1 change the second entry to the number of observations in your data file. Enter a title for the computer run in the second line. Enter the path to your data file on line 12. (If the data

are in the same directory as the program, just enter the file name.) Enter the format for the data on line 14 with a leading and a trailing parenthesis. (Note this is a FORTRAN representation of a data format.)

Figure 3-1 Toolkit Control File–toolkit.cnl

```
1.   1000    163   * num bootstraps; sample size
2.   toolkit run on gnp data
3.    0   * IBOTLAG * SMALLEST P CONSIDERED FOR PREWHITENING AR(P)
4.    10   * ITOPLAG * LARGEST  P CONSIDERED FOR PREWHITENING AR(P)
5.   922   * 0 < IDUM < 714026  INTEGER SEED FOR BOOTSTRAP GENERATOR
6.   y 0 0 0 0 0 0. 0.  0.40 0.  * bispectral test
7.   y 0 0 0 0 0 0. 0.  0.   0.  * bds test
8.   y 0 0 0 0 0 0. 0.  0.   0.  * bicovariance test
9.   y 0 0 0 0 0 0. 0.  0.   0.  * McLeod-Li test
10.  y 5 0 0 0 0 0. 0.  0.   0.  * Engle LM test
11.  y 5 0 0 0 0 0. 0.  0.   0.  * Tsay test
12.  gnp.dat
13.  1 0                   * start record-end record 0 means last rec
14.  (3x,f10.7)
```

Save the control file and exit the text editor or word processor. If you use a word processor, be sure the file is saved as an ASCII or text file.

3. To begin execution either double click on “toolkit.exe” in the Windows Explorer file listing or open a DOS box in Windows, go to the directory where the program is located, and type “toolkit” at a DOS prompt. (To open a DOS box, follow the sequence **Start...Programs...MS-DOS Prompt**.)

4. After the program has terminated, use a text editor or a word processor to examine the output file:

toolkit.out

HOW TO INTERPRET THE OUTPUT

The output file, toolkit.out, is too large to be easily displayed in a single figure. Therefore, we have used Figures 3-2 to 3-6 to display subsections of the output file.

Start with Figure 3-2. This is the portion of the file that follows the opening banner. Circled numbers 1 and 2 indicate the smallest and largest values of p that were considered for the AR(p) prewhitening model. The actual prewhitening model selected and applied is displayed farther down the file. Next follows a listing of descriptive statistics on the sample data. All of these are self explanatory, with the possible exception of C(6), which is the sample sixth-order cumulant of the data.

Figure 3-3 displays the last of the prewhitening models estimated and the actual selected prewhitening model. In the Figure, an AR(1) model was chosen because it minimized the Schwartz criterion for model selection.

Next the output file lists results on each test. The results labeled, "asymptotic theory" are based on the large sample distributions of the relevant test statistics, discussed in Chapter 2. For the bootstrap results, NBOOT "new" samples were independently drawn from the empirical distribution of the pre-whitened data. (NBOOT is set in the control file, discussed below; the "empirical distribution" of the data places weight 1/N on each of the N observed sample values.) Each new sample is used to calculate a value for the test statistic under the null hypothesis of serial independence. The fraction of the NBOOT test statistics thus obtained which exceed the sample value of the test statistic from the original data is then reported as the significance level at which the null hypothesis can be rejected.

Figure 3-4 lists the part of the output that reports on the Hinich Bispectral test and the McLeod-Li test. The Hinich Bispectral test gives the user the option of clipping the data, and applying a taper to the data in order to reduce the effect of side lobes on the spectral window[1]. These transformations are done before the test is done, and does not affect the other tests. The level of clipping is specified as some whole percentage. If the user had specified 5.0 as the clipping level in the control file, the highest 5% of the observations would have been clipped back to the 95% fractile, and the smallest 5% of the observations are pulled back to the 5% fractile. A modest amount of clipping, say 3 - 5% is recommended when the data is severely skewed. The taper parameter is expressed as a whole percentage also, and it represents the percentage of data points at the beginning and end of the series that are tapered with one-half a cosine bell. We recommend that, if tapering is used, the percentage be kept relatively small, about 10 - 15%. Both the clipping and tapering parameters were set to zero in the calculations reported here. The smoothing width, m, is the width of the square used to smooth the bispectrum in the frequency domain; the program sets m equal to the integer nearest to N^{1-e}. In order for the test statistics to converge asymptotically, the exponent e must be equal to or less than
0.50; a value of 0.40 was used in the calculations reported here. Four test statistics are reported for the Bispectral test. The interquartile range (IQR), the interdecile range (IDR), and the 80% fractile are the linearity test statistics. The tests for linearity are conditional on the rejection of Gaussianity for the data. In each case, what is reported

Figure 3-2 Toolkit Output File–Part 1

```
Version 1-27-99
Current Date: (MM/DD/YY) 01/28/99
Run at: (HH:MM:SS:hh) 13:47:37.14
title read in:
    toolkit run on gnp data

① SMALLEST P CONSIDERED FOR PREWHITENING AR(P) MODEL =  0
② LARGEST  P CONSIDERED FOR PREWHITENING AR(P) MODEL = 10
seed for bootstrap =   922

*********************************************************************
Total number obersevations on data file = 163
Number bootstrap iterations =1000

Sample data is read in from file gnp.dat

=====================================================================
      Descriptive Statistics of Data

   Sample size=      163

   Start observation from data set=        1

   Last observation used=      163
=====================================================================

Mean =  0.693E-02      Std Dev =  0.957E-02

Skew =-0.614       Kurtosis =  1.00       C(6) =    -4.027

Max value =  0.300E-01      Min value =  -0.260E-01
=====================================================================
```

Figure 3-3 Toolkit Output File–Part 2

Estimated AR(10) Model SC = -0.91160479E+01

Lag	Coefficient	T Ratio
1	0.28196	3.389
2	0.12151	1.419
3	-0.07156	-0.836
4	-0.01529	-0.178
5	-0.06535	-0.764
6	0.08912	1.043
7	0.00707	0.083
8	-0.15391	-1.814
9	-0.01354	-0.159
10	0.12362	1.526

Minimum SC AR(1) Model SC = -0.92936554E+01

Lag	Coefficient	T Ratio
1	0.31972	4.290

is the significance level at which the null hypothesis (of linearity or Gaussianity) can be rejected.

Shown below the Hinich Bispectral test in Figure 3-4 are the results for the McLeod-Li test. Significance levels are quoted up to the maximum lag, 24.

Figure 3-5 displays the output provided for the BDS test and the Bicovariance test. BDS results are calculated for embedding dimensions of 2, 3, and 4 and for ε equal to 0.50, 1.00 and 2.00 standard deviations. The results at $\varepsilon = 0.50$ are often erratic and those at $\varepsilon = 2.00$ are usually quite similar to those at $\varepsilon = 1.00$, so results at $\varepsilon = 1.00$ are reported below.

At the bottom of Figure 3-5 are the Bicovariance test results. The lag displayed is the maximum allowed for the estimator to converge asymptotically.

The part of the output file dealing with the Engle test and the Tsay test is shown in Figure 3-6. For the Engle test, the lags go from 1 to the highest lag requested in the control file.

As described in Chapter 2, the Tsay test makes use of lagged quadratic terms; the maximum lag considered is specified in the control file. Results are given for the overall test; estimated t ratios on the lagged quadratic terms are listed also.

Figure 3-4 Toolkit Output File–Part 3

```
HINICH BISPECTRAL TEST

percent clip = 0.0000
percent taper= 0.0000
Smoothing width, m =   21
Where m=n**(1-e)
Exponent e= 0.4000

bootstrap:

GAUSSIANITY    SIG. LEVEL = 0.345
IQR FRACTILE   SIG. LEVEL = 0.325
IDR FRACTILE   SIG. LEVEL = 0.331
80% FRACTILE   SIG. LEVEL = 0.269


asymptotic theory:

GAUSSIANITY    SIG. LEVEL = 0.017
IQR FRACTILE   SIG. LEVEL = 0.302
IDR FRACTILE   SIG. LEVEL = 0.572
80% FRACTILE   SIG. LEVEL = 0.161



MCLEOD-LI TEST

bootstrap:

USING UP TO :LAG  1     SIG. LEVEL =0.400
USING UP TO :LAG  2     SIG. LEVEL =0.228
USING UP TO :LAG  3     SIG. LEVEL =0.323
USING UP TO :LAG  4     SIG. LEVEL =0.418
USING UP TO :LAG  8     SIG. LEVEL =0.191
USING UP TO :LAG 12     SIG. LEVEL =0.119
USING UP TO :LAG 16     SIG. LEVEL =0.255
USING UP TO :LAG 20     SIG. LEVEL =0.292
USING UP TO :LAG 24     SIG. LEVEL =0.218


asymptotic theory:

USING UP TO :LAG  1     SIG. LEVEL =0.418
USING UP TO :LAG  2     SIG. LEVEL =0.311
USING UP TO :LAG  3     SIG. LEVEL =0.438
USING UP TO :LAG  4     SIG. LEVEL =0.530
USING UP TO :LAG  8     SIG. LEVEL =0.266
USING UP TO :LAG 12     SIG. LEVEL =0.156
USING UP TO :LAG 16     SIG. LEVEL =0.366
```

DETAILS OF THE CONTROL FILE

The control file, toolkit.cnl, contains the information the program needs to run the tests as desired by the user. All parameters and input data are read in free format; the user need not worry about what column each number starts or ends in – just separate each number by at least one space. All values are to be entered as integers (no decimal point) unless otherwise noted. We recommend that you use a text editor and merely modify the existing tooklit.cnl file in the distribution package. Looking at this file (or the example in Figure 3-1) you can see that comments can be added to each line after the data; similarly, an informative label can be added to the input data file after the last datum.

Line By Line Explanation

line 1: (NBOOT) Number of bootstrap simulations. A typical value for NBOOT is 1000; the maximum allowed value is 1000.
(N) Sample size – the number of observations to be read in from the data file. The maximum allowed value is 2000.

line 2: Title for job; only the first 75 characters will be used.

line 3: (IBOTLAG) The data is prewhitened using an AR(p) model. This parameter gives the smallest value of p considered; IBOTLAG must be nonnegative.

line 4: (ITOPLAG) The data is prewhitened using an AR(p) model. This parameter gives the largest value of p considered; ITOPLAG must be in the interval [IBOTLAG, 16]. Choosing ITOPLAG equal to IBOTLAG allows the user to specify a value for p.

line 5: (IDUM) Starting seed for the random number used in generating the bootstrapped samples. IDUM must lie in the interval [0, 714026].

line 6: Control line for the Hinich Bispectral test. The first item on this line must be a lower class "y" to specify that the test will be done, or a lower case "n" to specify that the test not be done. Then enter 5 zeroes, each separated by at least one space. Enter in the next two fields (1) percent clipping desired (0.0 for no clipping) and (2) the percent taper desired (0.0 for no taper). The clip and taper parameter are entered as percents, not decimal fractions. In the ninth field enter a value for "e" with a decimal point, then another zero with a decimal point. The smoothing parameter M defined in Chapter 2 will be set to the integer nearest to $N^{(1-e)}$; a typical value for e would be 0.40.

line 7: Control line for the BDS test. The first item on this line must be a lower case "y" to specify that the test will be done or a lower case "n" to specify that the test not be done. Then enter 5 zeroes, each separated by at least one space, then four zeroes with decimal points.

line 8: Control line for the Bicovariance test. It is filled out the same as line 7.

line 9: Control line for the McLeod-Li test. It is filled out the same as line 7.

line 10: Control line for the Engle LM test. The first item on this line must be a lower class "y" to specify that the test will be done, or a lower case "n" to specify that the test not be done. The next item is an integer specifying how many lags to use in the test. A typical value would be 5; the maximum allowed value is 50. Then enter 4 zeroes, each separated by at least one space, and 4 zeroes with decimal points.

line 11: Same as line 10 only for the Tsay test; the maximum number of lags allowed is 12.

Figure 3-5 Toolkit Output File–Part 4

BDS TEST SIGNIFICANCE LEVELS

bootstrap:

DIMENSION	EPS = 0.50	EPS = 1.00	EPS = 2.00
2	0.354	0.356	0.197
3	0.223	0.041	0.041
4	0.349	0.031	0.028

asymptotic theory:

DIMENSION	EPS = 0.50	EPS = 1.00	EPS = 2.00
2	0.362	0.393	0.177
3	0.174	0.024	0.022
4	0.347	0.010	0.009

BICOVARIANCE TEST

bootstrap:

USING UP TO LAG: 7 SIG. LEVEL =0.017

asymptotic theory:

USING UP TO LAG: 7 SIG. LEVEL =0.008

Figure 3-6 Toolkit Output File–Part 5

ENGLE TEST

bootstrap:

USING UP TO LAG: 1 SIG. LEVEL =0.367
USING UP TO LAG: 2 SIG. LEVEL =0.205
USING UP TO LAG: 3 SIG. LEVEL =0.267
USING UP TO LAG: 4 SIG. LEVEL =0.372
USING UP TO LAG: 5 SIG. LEVEL =0.525

asymptotic theory:

USING UP TO LAG: 1 SIG. LEVEL =0.397
USING UP TO LAG: 2 SIG. LEVEL =0.279
USING UP TO LAG: 3 SIG. LEVEL =0.366
USING UP TO LAG: 4 SIG. LEVEL =0.475
USING UP TO LAG: 5 SIG. LEVEL =0.609

TSAY TEST SIGNIFICANCE LEVELS

bootstrap : 0.025

asymptotic theory: 0.029

Average t ratios from Tsay regression:

lag	lag	t ratio
1	1	0.114
2	1	0.468
2	2	1.602
3	1	-0.643
3	2	2.102
3	3	0.273
4	1	-1.391
4	2	0.711
4	3	1.105
4	4	-0.132
5	1	1.385
5	2	-2.969
5	3	1.966
5	4	-0.773

line 12: Name of the file containing the data. The user must provide the complete path unless the file is located on the same directory as the "toolkit.exe," the executable file. If this entry is set to "*" (omitting the quotation marks), then the default file name ("toolkit.dat") is assumed.

line 13: The first entry is the first observation to use in the analysis. This is followed by an entry specifying the last observation to use. If this second entry is set to zero, then the analysis will include the last observation available.

line 14: The format for the input data. This is entered as a FORTRAN format. The data is read in free format if this entry is set to "*".

References:

Jenkins, G. and D. Watts (1968) *Spectral Analysis and Its Applications* Holden-Day: San Francisco.

Notes:

1. A discussion of spectral windows and the side lobes induced by a rectangular data window can be found in chapters 6 and 7 of Jenkins and Watts (1968).

4 ARTIFICIALLY GENERATED DATA: SIZE CONSIDERATIONS

INTRODUCTION

The size of a statistical test is the probability with which it wrongly rejects the null hypothesis. Thus, for data which is actually generated by a linear mechanism, a 5% nonlinearity test – a test whose size is .05 – will reject the null hypothesis of linearity 5% of the time.[1] However, like most econometric procedures, the nonlinearity tests described above in Chapter 2 are only asymptotically justified. For a 5% test, this implies that the actual size of the test is known to approach .05 as the sample length becomes arbitrarily large, but it says little about the size of the test for any given, finite sample length.

In most applications a sample length of several hundred observations is sufficiently long that the actual size closely approximates the asymptotic (nominal) size. This rule of thumb is suspect in analyzing nonlinear time series processes, however. Intuitively, this follows from the fact that testing for nonlinearity often involves a consideration of the higher order moments of a process, whereas it is well known that accurate estimation of higher moments requires larger samples.

Particular concern has been expressed in the literature about the validity of the BDS test for reasonable sample lengths and addressed, to some degree, in Brock, et al. (1991). More recently, de Lima (1997) has considered the behavior of a number of nonlinearity tests where the moment restriction assumptions underlying the asymptotic distributions of these tests are not satisfied, finding particular problems in situations involving leptokurtic (heavy-tailed) data.

METHODOLOGY

Because we share these concerns, we routinely bootstrap the significance levels of all the tests used here, as well as computing significance levels based on asymptotic theory. This application of the bootstrap is very straightforward. As described in Chapter 2, the N sample observations are first prewhitened using a linear autoregressive model, so that the resulting data is (asymptotically) serially i.i.d. under the null hypothesis of a linear generating mechanism. We then draw 1000 separate samples of length N at random from the above prewhitened series.[2] The bootstrap significance level for a given test is then just the fraction of these 1000 "new" N-samples for which the test statistic exceeds that observed in the (prewhitened) sample data. It is simple enough to confirm that 1000 bootstrap replications is sufficient by merely observing that the results are invariant to increasing this number; it is distinctly less clear that N itself is sufficiently large: after all, both the prewhitening procedure and the bootstrap itself are themselves only asymptotically justified.[3]

Consequently, we used N_{trial} Monte Carlo simulations to examine the actual size of each test, using both asymptotic theory and the bootstrap, with samples of length 200. Since the effect of prewhitening is an issue, we generated both serially i.i.d. variates

$$x_t = \varepsilon_t$$

and variates with substantial linear serial dependence induced by the AR(2) model:

$$x_t = .4x_{t-1} + .3x_{t-2} + \varepsilon_t$$

where the ε_t are independently generated from each of four distributions: Gaussian, exponential, Student's *t* with 5 degrees of freedom, and the symmetric stable Paretian distribution with $\alpha = 1.93$[4]. Histograms of the empirical frequency distribution of 10,000 realizations of ε_t drawn from each of these distributions are given in Figures 4-1 through 4-4. The exponential distribution is quite asymmetric. Both of the latter two distributions are heavy-tailed – to the point where the symmetric stable Paretian distribution with this index value has infinite variance.

Since the null hypothesis of a linear generating mechanism is satisfied by construction for these simulated data, the actual size of each test is estimated by the fraction of N_{trial} generated N-samples for which the test (incorrectly) rejects the null hypothesis. Each test is performed at the 5% significance level. Therefore, if N is sufficiently large, the empirical size ought to be close to .05 for each test.

Recalling that N_{trial} is the number of Monte Carlo trials, the empirical size is asymptotically distributed $N[p, p(1-p)/N_{trial}]$ under the assumption that the true size of the test is p. Taking $N_{trial} = 1000$ to be sufficiently large that this asymptotic distribution is a good approximation to the actual distribution, the empirical size should lie within the interval [.036, .064] with probability .95 if the true size of the test is .05.[5]

Figure 4-1 Histogram of 10, 000 Gaussian Variates

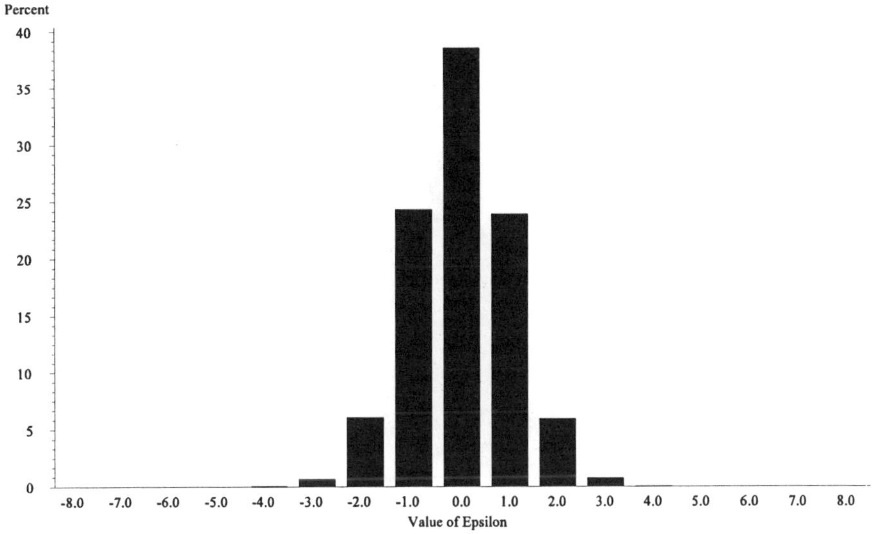

Figure 4-2 Histogram of 10, 000 Exponential Variates

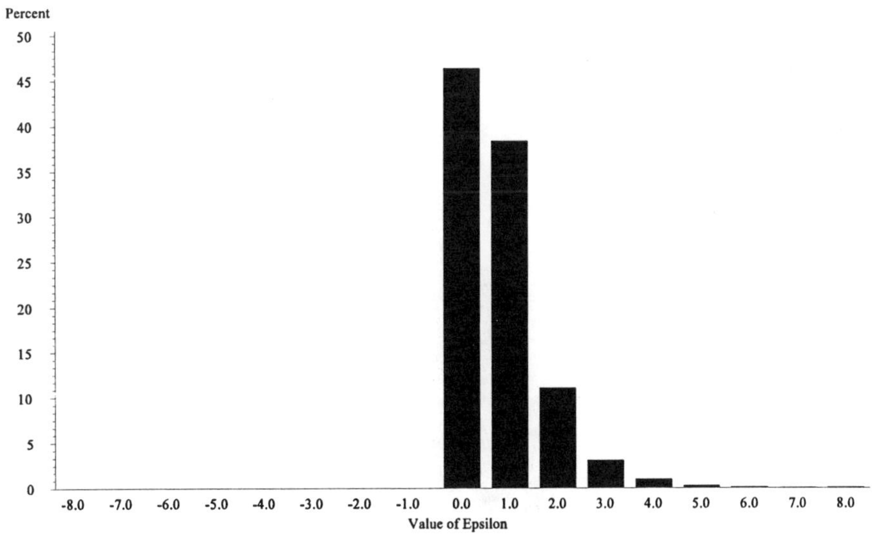

Figure 4-3 Histogram of 10, 000 Student's t Variates

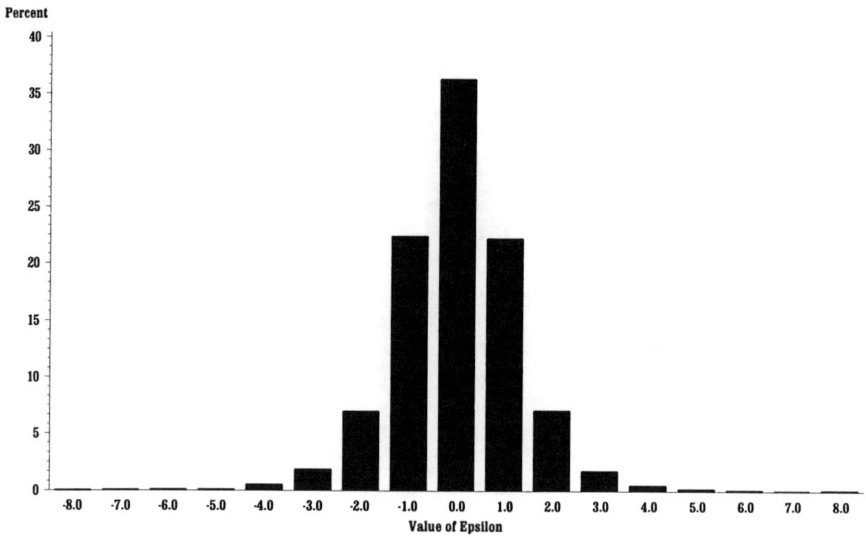

Figure 4-4 Histogram of 10, 000 Symmetric Pareto Variates

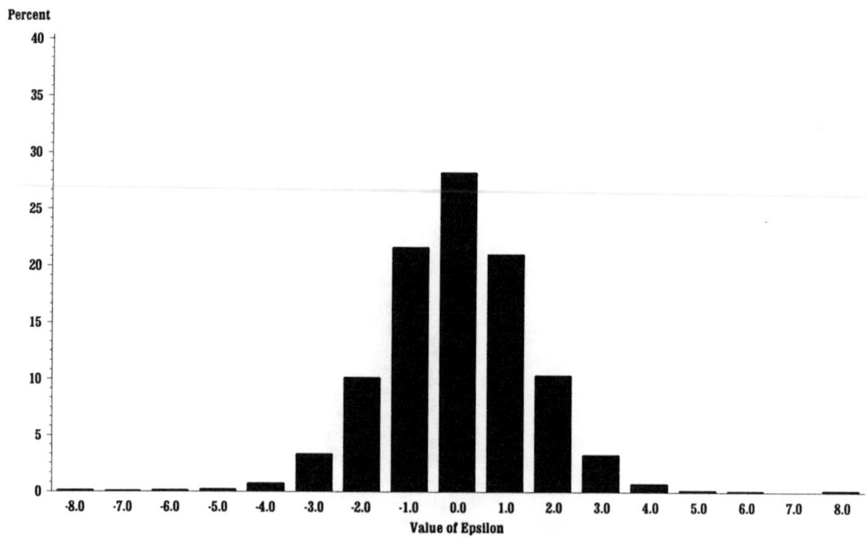

RESULTS

The results of these calculations for N = 200 are given in Table 4-1 for the serially i.i.d. variates and in Table 4-2 for the AR(2) variates; results of calculations for the AR(2) variates with N = 500 are given in Table 4-3.

In Table 4-1, for the i.i.d. variates with N = 200, we observe that the sizes of the bootstrapped tests are essentially correct at this sample length, regardless of the form of the distribution from which the data were drawn – the fact that two of the size estimates for bootstrapped tests lie outside the 95% confidence interval around .05 is inconsequential in view of the number of estimates made. In contrast, our Monte Carlo estimates of the sizes of the tests based on asymptotic theory – in particular, the BDS and bispectral tests – are typically a bit high.

In Table 4-2, for the AR(2) variates with N = 200, we observe three bootstrapped tests with size estimates outside the 95% interval of [.036, .064]; this is a bit more than one might expect due to chance alone. This minor discrepancy is due to sampling errors in identifying and estimating the prewhitening model using only 200 observations. Turning to the Monte Carlo estimates of the sizes of the tests based on asymptotic theory, the sizes of the BDS tests are still typically a bit high and those of the bispectral test can be too high or quite a bit too low, depending on the distribution. Apparently, N = 200 is too small a sample length for the asymptotic results to be reliable using either the i.i.d. variates or the AR(2) variates.

Table 4-3 reports results for the AR(2) variates at N = 500. Here we observe that all of the size estimates of the bootstrapped tests are within the 95% interval of [.036, .064]; apparently, the prewhitening procedure works reasonably well at N = 500. With the exception of the bispectral test, most of the tests based on asymptotic theory work better at N = 500 than at N = 200, but there are still more size estimates which differ significantly from .05 than one would expect from chance alone. For the exponential and Paretian distributions the performance of the bispectral test based on asymptotic theory is notably worse at N = 500 than it was at N = 200; apparently, this test must be bootstrapped even with fairly large samples.

We conclude that the actual sizes of the bootstrapped tests are reasonably satisfactory using N ≥ 200 observations {either i.i.d. or AR(2)} generated from any one of the four distributions. In contrast, the tests based on asymptotic theory appear to require sample lengths in excess of 200 observations, in some cases in excess of 500 observations. Consequently, we proceed using the bootstrapped tests for samples of roughly 200 observations in length or longer without further concern about prewhitening or the form of the data's distribution.

Table 4-1
Empirical Size of 5% Tests[a]
Serially i.i.d. Data – 200 Observations

	McLeod-Li	Engle LM	BDS			Tsay	Bicov.	Bispectral
	L = 24	p = 5	m = 2	m = 3	m = 4	k = 5	ℓ = 8	M = 24
Bootstrap								
Gaussian	.055	.057	.051	.048	.054	.048	.061	.040
Student's *t*	.048	.055	.050	.043	.044	.045	.051	**.034***
Exponential	.052	.055	.062	.057	.058	.064	.055	.050
Paretian	**.028***	.039	.053	.051	.057	.045	.052	.045
Asymptotic Theory								
Gaussian	.055	.050	**.074***	**.081***	**.079***	.044	.059	**.065***
Student's *t*	.049	.045	**.070***	**.068***	**.074***	.040	.055	**.006***
Exponential	.062	**.066***	**.067***	**.074***	**.072***	**.074***	**.081***	**.101***
Paretian	**.034***	**.032***	**.068***	**.075***	**.078***	.048	.058	.050

[a] Results outside the interval [.036, 064] are significantly different from .050 and marked with an asterisk; see text for details. The parameters L, p, m, k, ℓ , and M are defined in Chapter 2, where each test is discussed; as indicated there, the Hinich bispectral test is implemented using the interdecile range.

Table 4-2
Empirical Size of 5% Tests[a]
AR(2) Generating Process – 200 Observations

	McLeod-Li	Engle LM	BDS			Tsay	Bicov.	Bispectral
	L = 24	p = 5	m = 2	m = 3	m = 4	k = 5	ℓ = 8	M = 24
Bootstrap								
Gaussian	.058	.060	.045	.049	.053	.045	.040	.047
Student's *t*	.041	.050	**.032***	.041	.041	.052	.060	.052
Exponential	.037	.043	.051	.053	.056	**.068***	.055	.054
Paretian	**.034***	.047	.048	.038	.040	.058	.045	.041
Asymptotic Theory								
Gaussian	.063	.047	**.073***	**.076***	**.073***	.037	.041	**.013***
Student's *t*	.048	.039	.052	**.066***	**.068***	.044	.060	**.021***
Exponential	.051	.056	.059	**.068***	**.067***	**.078***	**.080***	**.078***
Paretian	.043	.040	**.065***	.056	.058	.058	.051	.058

[a] Results outside the interval [.036, 064] are significantly different from .050 and marked with an asterisk; see text for details. The parameters L, p, m, k, ℓ , and M are defined in Chapter 2, where each test is discussed; as indicated there, the Hinich bispectral test is implemented using the interdecile range.

Table 4-3
Empirical Size of 5% Tests[a]
AR(2) Generating Process – 500 Observations

	McLeod-Li	Engle LM	BDS			Tsay	Bicov.	Bispectral
	L = 24	p = 5	m = 2	m = 3	m = 4	k = 5	ℓ = 11	M = 42
Bootstrap								
Gaussian	.049	.051	.047	.052	.047	.052	.042	.054
Student's *t*	.053	.059	.043	.042	.045	.052	.061	.040
Exponential	.045	.055	.050	.055	.055	.056	.048	.051
Paretian	.046	.037	.056	.052	.042	.057	.055	.043
Asymptotic Theory								
Gaussian	.052	.046	.057	.060	.058	.047	.046	**.067***
Student's *t*	.056	.058	.049	.045	.056	.047	.064	.059
Exponential	**.081***	**.076***	.050	.057	.057	.063	**.105***	**.253***
Paretian	.054	.040	.060	.061	.053	.061	**.076***	**.309***

[a] Results outside the interval [.036, 064] are significantly different from .050 and marked with an asterisk; see text for details. The parameters L, p, m, k, ℓ, and M are defined in Chapter 2, where each test is discussed; as indicated there, the Hinich bispectral test is implemented using the interdecile range.

REFERENCES

Brock, W. A., Hsieh, D. A., and LeBaron, B.D. (1991) *A Test of Nonlinear Dynamics, Chaos, and Instability: Theory and Evidence* MIT Press: Cambridge.

Fama, E. F. (1965) "The Behavior of Stock Market Prices" *Journal of Business 38*, 34-105.

Efron, B (1979) "Bootstrap Methods: Another Look at the Jackknife." *Annals of Statistics 7*, 1-26.

Kanter, M. and Steiger W. L. (1974) "Regression and Autoregression with Infinite Variance" *Advances in Applied Probability 6*, 768-83.

de Lima, P. J. F. (1997) "On the Robustness of Nonlinearity Tests to Moment Condition Failure" *Journal of Econometrics 76*, 251-80.

NOTES

1.The size of a test is synonymous with its "significance level." That is, when a 5% test rejects the null hypothesis, one can conclude that "linearity was rejected at the 5% level," etc.

2.Each of these 1000N draws is obtained in the same way: An integer (k) in the interval [1, N] is chosen at random. The "new" draw is then set equal to the kth (prewhitened) sample observation. The random integers are drawn using a shuffled linear congruential generator. Because the bootstrap requires prewhitening, in the bootstrapped implementation of the Hinich bispectral test, the denominator of Ψ^2 in Theorem 2-1 is replaced by the cube of the sample variance of the prewhitened data.

3.E.g., see Efron (1979).

4.Symmetric stable Paretian variates were simulated using the exact algorithm given by Kanter and Steiger (1974). The de Lima (1997) article arbitrarily considers $\alpha = 1.50$; we chose $\alpha = 1.93$ because this is the value Fama (1965) estimates for U.S. stock price data.

5.The value of N_{trial} bears no particular relationship to the number of bootstrap replications used in the bootstapped tests. Here it is set to a fairly large number so that we can obtain relatively precise estimates of the sizes of the tests. In later chapters, where we use Monte Carlo simulations to estimate the power of various tests, N_{trial} is set to a lower value (250) because this degree of precision is not necessary.

5 ARTIFICIALLY GENERATED DATA: POWER AND MODEL SPECIFICATION CONSIDERATIONS

INTRODUCTION

In this Chapter we discuss our estimates of the power of each test against a number of alternative data generating processes.[1] Our goal is to answer the following questions:
1. Do one or more of the tests have high power against *all* of the alternative processes?
2. Is the *pattern* of power estimates across the alternative processes similar for all of the tests?
3. For a given generating process, are the results of all of the tests highly correlated across the simulations?

If one of the tests dominates all the rest in terms of power, then this test is the one to use as a "nonlinearity screening test" to, for example, routinely check the fitting errors of a proposed model for a time series. On the other hand, since such a test has relatively high power against *all* of the alternatives, it conveys very little information as to which kind of nonlinear model is appropriate.

Paradoxically, such identifying information is only obtainable from tests whose performance is uneven across the alternatives. In this context, a test conveys identifying information to the extent that it is either particularly powerful *or* particularly *un*powerful against a limited subset of the alternatives.

Finally, we examine the correlations between the results of the tests for a given alternative. Here, again paradoxically, what is useful is a lack of consistency. For example, suppose that we generate 250 samples of length T from a particular nonlinear process. If test #1 rejects the null hypothesis of linearity at the 5% level for, say, 220 of these samples, then it has higher power than test #2 which only rejects the null for 190 samples. But were most of these 190 samples among the 220 samples for which test #1 rejected, or not? If both tests reject over basically the same samples, then test #2 is simply an inferior alternative to test #1 for data generated from this particular process. In contrast, if this "rejection overlap" is small, then test #2 is sensitive to a different aspect of the data set than test #1 and provides separately useful information as to whether or not to reject the null hypothesis – in this case it is worthwhile to do both tests and perhaps combine them into a single joint test.

GENERATING PROCESSES

The space of all possible nonlinear processes is incredibly broad. The literature in the field contains many, many examples of nonlinear processes. Not only does each paper in the literature have its own list of nonlinear processes, but there is no guarantee that even the union of these lists spans the space in any meaningful sense.[2] The eight generating models considered here are given in Table 5-1. We make no representation that this list even spans the space of all models considered in the literature, but it does include a number of different nonlinear models, each of which has either been applied empirically or corresponds to the low order terms in a Volterra expansion of a rather general class of nonlinear models.[3]

The first nonlinear processes listed are ARCH and GARCH processes, as introduced by Engle (1982) and Bollerslev (1986); the ARCH process used here corresponds to Engle's preferred model for the U.K. inflation rate. Although many nonlinear processes display conditional heteroskedasticity, the ARCH(4) and GARCH(1,1) models simulated here (like all ARCH/GARCH processes) are "pure" in that they are martingale difference processes: by construction, only their variance is serially dependent.

The next process listed is a generic switching regression. A number of switching models have been considered in the literature – e.g., Tong (1983), Hamilton (1989), and Teräsvirta and Anderson (1992). Here, we have simulated a typical threshold autoregressive (TAR) model from Tong and Lim (1980) and a simple Markov switching model in the spirit of Hamilton (1989).[4] Finally, we consider several models suggested by the usual Volterra expansion. Two quadratic models are considered, so as to include both a forecastable process and a martingale process. Two analogous cubic models are also considered.

Each of the data generation models was simulated using serially i.i.d innovations. These innovations were generated from the

- unit normal distribution,
- Student's *t* distribution with 5 degrees of freedom (to simulate the effects of a leptokurtic (fat-tailed) distribution),
- exponential distribution (to simulate the effects of an asymmetric innovation distribution),

and

- symmetric stable Paretian distribution, to simulate the effect of an innovation distribution for which finite second (and higher) moments do not exist.[5]

Table 5-1 Summary of Data Generating Processes Considered[a]

Conditional Heteroskedasticity Models:	
ARCH(4)	$x_t = (h_t)^{.5}\,\varepsilon_t$ $h_t = .000019 + .846\{x^2_{t-1} + .3x^2_{t-2} + .2x^2_{t-3} + .1x^2_{t-4}\}$
GARCH(1,1)	$x_t = (h_t)^{.5}\,\varepsilon_t$ $h_t = .011 + .12\,(x_{t-1})^2 + .85\,h_{t-1}$
Switching Models:	
TAR	$x_t = -.5\,x_{t-1} + \varepsilon_t$ if $x_{t-1} < 1$ $x_t = .4\,x_{t-1} + \varepsilon_t$ otherwise
Two State Markov	$x_t = -.5\,x_{t-1} + \varepsilon_t$ if in state 1 $x_t = .4\,x_{t-1} + \varepsilon_t$ if in state 2 (Remain in state with probability .90)
Quadratic models:	
martingale	$x_t = \varepsilon_t + .8\,\varepsilon_t\,[\varepsilon_{t-1} + .8\varepsilon_{t-2} + .8^2\varepsilon_{t-3} + \ldots + .8^{17}\varepsilon_{t-18}]$
non-martingale	$x_t = \varepsilon_t + .8\,\varepsilon_{t-1}\,[\varepsilon_{t-2} + .8\varepsilon_{t-3} + .8^2\varepsilon_{t-4} + \ldots + .8^{17}\varepsilon_{t-19}]$
Cubic models:	
martingale	$x_t = \varepsilon_t + .8\,\varepsilon_t\,[\varepsilon_{t-1}\varepsilon_{t-2} + .8\varepsilon_{t-2}\varepsilon_{t-3} + \ldots + .8^{17}\varepsilon_{t-18}\varepsilon_{t-19}]$
non-martingale	$x_t = \varepsilon_t + .8\,\varepsilon_{t-1}\,[\varepsilon_{t-2}\varepsilon_{t-3} + .8\varepsilon_{t-3}\varepsilon_{t-4} + \ldots + .8^{17}\varepsilon_{t-19}\varepsilon_{t-20}]$

[a] ε_t is iid(0,1) in all models. As noted in the text, the distributions for ε_t considered were: Gaussian, Student's *t*, exponential, and symmetric Paretian.

POWER RESULTS

Our estimated power results are given in Tables 5-2a,b and 5-3a-h below for each of the tests discussed in Chapter 2. Tables 5-2a,b summarize the results for all eight generating processes in Table 5-1, driven by Gaussian innovations, with sample sizes of 200 and 500. Each of Tables 5-3a-h focuses on one generating process at N = 200 and compares the power of the tests across the four innovation distributions considered.

No single test dominates all the others across all eight alternative generating processes. However, the BDS test clearly stands out in terms of overall power against a variety of alternatives. It has distinctly the highest power for the ARCH, Markov switching, martingale quadratic, and for both cubic processes; it also does quite well against the GARCH and the non-martingale quadratic processes. For the TAR process the Tsay test stands out, but even there the BDS test still exhibits reasonable power. At N = 500 (Table 5-2b) and using the other three distributions (Tables 5-3a-h) we find basically the same pattern of results. We conclude that the BDS test is the best test of this group for use as a "nonlinearity screening test."

On the other hand, this same consistently high power across the alternatives also implies that the BDS test conveys very little information as to what kind of nonlinear process generated the data. Here it is *in*consistent power against the alternatives that is useful. In this context the Tsay test stands out as a possible marker for TAR models. The results given in Table 5-3c indicate that this result holds up across all four innovation distributions but, obviously, this result needs to be confirmed across a variety of different TAR and other switching models.[6]

Table 5-2a Power Estimates of 5% Tests[a]
Gaussian Innovations – 200 Observations

	McLeod-Li	Engle LM	BDS			Tsay	Bicov.	Bispectral
	L = 24	p = 5	m = 2	m = 3	m = 4	k = 5	$\ell = 8$	M = 24
Conditional Heteroskedasticity Models								
ARCH	.78	.91	1.00	1.00	1.00	.42	.60	.16
GARCH	.75	.72	.63	.76	.82	.43	.74	.14
Switching Models								
TAR	.10	.18	.65	.66	.60	.84	.13	.17
Markov	.16	.33	.57	.54	.51	.07	.15	.04
Quadratic Models								
martingale	.45	.64	.96	.98	.99	.56	.67	.13
non-martingale	.04	.11	.16	.23	.26	.28	.29	.06
Cubic Models								
martingale	.78	.89	.99	1.00	1.00	.84	.96	.22
non-martingale	.56	.72	.88	.94	.96	.57	.72	.12

[a]Based on 250 generated samples and 1000 bootstrap replications. L, p, $m\ell$, k, and M are defined in Chapter 2, where each test is discussed; as indicated there, the Hinich bispectral test is implemented using the interdecile range. For BDS tests ε equals one standard deviation.

Table 5-2b Power Estimates of 5% Tests[a]
Gaussian Innovations – 500 Observations

	McLeod-Li	Engle LM	BDS			Tsay	Bicov.	Bispectral
	L = 24	p = 5	m = 2	m = 3	m = 4	k = 5	ℓ = 8	M = 24
Conditional Heteroskedasticity Models								
ARCH	.98	1.00	1.00	1.00	1.00	.53	.78	.22
GARCH	.99	.99	.95	.98	.99	.61	.98	.33
Switching Models								
TAR	.16	.37	.95	.91	.87	1.00	.14	.38
Markov	.47	.73	.92	.86	.79	.10	.16	.08
Quadratic Models								
martingale	.86	.96	1.00	1.00	1.00	.68	.86	.16
non-martingale	.90	.98	1.00	1.00	1.00	.86	.94	.36
Cubic Models								
martingale	.97	.98	1.00	1.00	1.00	.98	1.00	.24
non-martingale	.90	.98	1.00	1.00	1.00	.72	.94	.22

[a]Based on 250 generated samples and 1000 bootstrap replications. L, p, mℓ , k, and M are defined in Chapter 2, where each test is discussed. For BDS tests ε equals one standard deviation.

Table 5-3a Empirical Power of 5% Tests[a]
Data generated from ARCH(4) model[b] – 200 Observations

	McLeod-Li	Engle LM	BDS			Tsay	Bicovariance	Bispectral
	$L = 24$	$p = 5$	$m = 2$	$m = 3$	$m = 4$	$k = 5$	$\ell = 5$	$M = 24$
Gaussian	.78	.91	1.00	1.00	1.00	.42	.60	.16
Student's *t*	.76	.89	.98	1.00	1.00	.39	.58	.16
Exponential	.24	.36	.92	.98	.99	.63	.60	.22
Paretian	.54	.74	1.00	1.00	1.00	.54	.67	.27

[a]Based on 250 generated samples and 1000 bootstrap replications. L, p, m, ℓ, k, and M are defined in Chapter 2, where each test is discussed. For BDS tests ε equals one standard deviation.

[b] See Table 5-1.

Table 5-3b
Empirical Power of 5% Tests
Data generated from GARCH(1,1) model – 200 Observations

	McLeod-Li	Engle LM	BDS			Tsay	Bicovariance	Bispectral
	L = 24	p = 5	m = 2	m = 3	m = 4	k = 5	ℓ = 5	M = 24
Gaussian	.75	.72	.63	.76	.82	.43	.74	.32
Student's *t*	.47	.60	.57	.68	.77	.23	.46	.19
Exponential	.78	.71	.93	.97	.98	.79	.93	.11
Paretian	.91	.94	.97	.98	.98	.97	.98	.29

Table 5-3c
Empirical Power of 5% Tests
Data generated from TAR model – 200 Observations

	McLeod-Li	Engle LM	BDS			Tsay	Bicovariance	Bispectral
	L = 24	p = 5	m = 2	m = 3	m = 4	k = 5	ℓ = 8	M = 24
Gaussian	.13	.16	.71	.66	.60	.85	.14	.19
Student's *t*	.07	.16	.67	.61	.59	.85	.08	.21
Exponential	.08	.14	.92	.97	.98	.83	.15	.18
Paretian	.06	.16	.57	.49	.41	.66	.16	.17

Table 5-3d
Empirical Power of 5% Tests
Data generated from Two State Markov model – 200 Observations

	McLeod-Li	Engle LM	BDS			Tsay	Bicov.	Bispectral
	L = 24	p = 5	m = 2	m = 3	m = 4	k = 5	ℓ = 8	M = 24
Gaussian	.16	.33	.57	.54	.51	.07	.15	.04
Student's *t*	.17	.40	.58	.56	.50	.10	.16	.05
Exponential	.24	.44	.89	.92	.89	.30	.36	.09
Paretian	.15	.32	.70	.65	.60	.14	.24	.07

Table 5-3e
Empirical Power of 5% Tests
Data Generated from Quadratic Martingale Model – 200 Observations

	McLeod-Li	Engle LM	BDS			Tsay	Bicovariance	Bispectral
	L = 24	p = 5	m = 2	m = 3	m = 4	k = 5	$\ell = 8$	M = 24
Gaussian	.45	.64	.96	.98	.99	.56	.67	.13
Student's *t*	.44	.64	.96	.99	1.00	.54	.69	.21
Exponential	.12	.23	.49	.61	.68	.31	.40	.12
Paretian	.40	.60	.98	.99	.99	.64	.67	.21

Table 5-3f
Empirical Power of 5% Tests
Data Generated from Quadratic Non-martingale Model – 200 Observations

	McLeod-Li	Engle LM	BDS			Tsay	Bicovariance	Bispectral
	L = 24	p = 5	m = 2	m = 3	m = 4	k = 5	$\ell = 8$	M = 24
Gaussian	.04	.11	.16	.23	.26	.28	.29	.06
Student's *t*	.52	.69	.92	.98	.98	.56	.70	.17
Exponential	.51	.76	.99	1.00	1.00	.91	.91	.15
Paretian	.52	.69	.99	.99	1.00	.63	.74	.22

Table 5-3g
Empirical Power of 5% Tests
Data Generated from Cubic Martingale model – 200 Observations

	McLeod-Li	Engle LM	BDS			Tsay	Bicovariance	Bispectral
	L = 24	p = 5	m = 2	m = 3	m = 4	k = 5	ℓ = 8	M = 24
Gaussian	.78	.79	.99	1.00	1.00	.84	.96	.22
Student's *t*	.84	.74	.97	1.00	1.00	.79	.94	.14
Exponential	.29	.27	.60	.71	.80	.48	.61	.10
Paretian	.58	.66	.96	.98	.98	.76	.90	.20

Table 5-3h
Empirical Power of 5% Tests
Data Generated from Cubic Non-martingale model – 200 Observations

	McLeod-Li	Engle LM	BDS			Tsay	Bicovariance	Bispectral
	L = 24	p = 5	m = 2	m = 3	m = 4	k = 5	ℓ = 8	M = 24
Gaussian	.56	.72	.88	.94	.96	.57	.72	.12
Student's *t*	.57	.74	.85	.95	.96	.56	.72	.16
Exponential	.58	.78	1.00	1.00	1.00	.98	.93	.17
Paretian	.58	.72	.98	1.00	1.00	.76	.88	.18

Next we turn to the third question raised at the beginning of this chapter. Generally speaking, there appear to be few complementarities among the tests – when a particular test's power exceeds that of another against a given alternative, we typically find that the less powerful test is rejecting the null hypothesis over basically the same sample replications in which the more powerful test rejects the null also. For example, a crosstabulation is given in Table 5-4 below of the BDS vs. Tsay test results for the TAR and cubic non-martingale processes driven by Gaussian innovations. The Tsay test has higher power than the BDS test for the TAR alternative; the BDS test has higher power against the cubic non-martingale alternative. In both cases, however, the less powerful test rejects on only a few percent of the simulated data sets for which the more powerful test fails to reject.

This result is particularly interesting in view of the fact that the (transformed) test results are not highly correlated across the 250 samples drawn from each generating process: this sample correlation is .54 for data generated from the TAR process and only .30 for the results on data generated from cubic non-martingale process. Each bootstrap significance level is the fraction of the test statistics generated from 1000 samples bootstrapped from the (prewhitened) data {generated from a particular– e.g., TAR – model} which fail to exceed the sample value for that data set. Thus, we obtain 250 significance levels for each test. The "transformation" referred to above is just the inverse truncated Gaussian distribution function. This transformation "spreads out" the 250 significance levels by making them approximately Gaussian; this makes the scatterplots – Figure 5-1 and 5-2 below – substantially more interpretable.[7]

Table 5-4
BDS Test Versus Tsay Test Crosstabulation

	Generating Model[a]	
	TAR	Cubic
Power for 5% BDS Test	.66	.94
Power for 5%Tsay Test	.84	.57
Fraction of replications both tests reject null	.61	.55
Fraction BDS test alone rejects	.05	.39
Fraction Tsay test alone rejects	.23	.01
Fraction neither test rejects	.10	.05
Sample correlation between test results[b]	.54	.30

[a] "Cubic" refers to the cubic non-martingale generating model from Table 5-1; both models are driven by Gaussian innovations.

[B] Sample correlation between the 250 transformed significance levels for the indicated tests, obtained from 250 samples (each of length 200 observations) generated using the indicated processes. As described in Chapter 2, each significance level is obtained using 1000 bootstrap simulations, each of which is drawn from the residuals of a prewhitening model for the sample data. Each significance level is transformed using an inverse truncated Gaussian c.d.f. so as to make Figures 5-1 and 5-2 more interpretable. Both sample correlations appear to be significantly positive based on the usual t test, but this test assumes Gaussianity, an assumption which the crossplots (Figures 5-1 and 5-2) indicate is still violated even for the transformed significance levels. (This violation occurs because "only" 1000 bootstrap simulations were used.)

Figure 5-1 is a scatterplot of the transformed significance levels for the two tests using data generated from TAR model given in Table 5-1; Figure 5-2 gives an analogous scatterplot of the results obtained using data generated from the non-martingale cubic model. It is apparent from these plots that the sample correlations given in Table 5-4 are in fact overstating the degree of relatedness between the results for these two tests, particularly for the data generated from the cubic non-martingale process. Nevertheless, the potential for power gains from combining these tests is necessarily severely limited by the paucity of simulations in which the less powerful test rejects the null hypothesis when the other test does not. Consequently, we conclude that construction of combination tests is probably not worth pursuing.

Figure 5-1 Cross-Plot of Threshold Autoregressive Model Test Results:
Transformed Significance levels for Tsay and BDS Tests

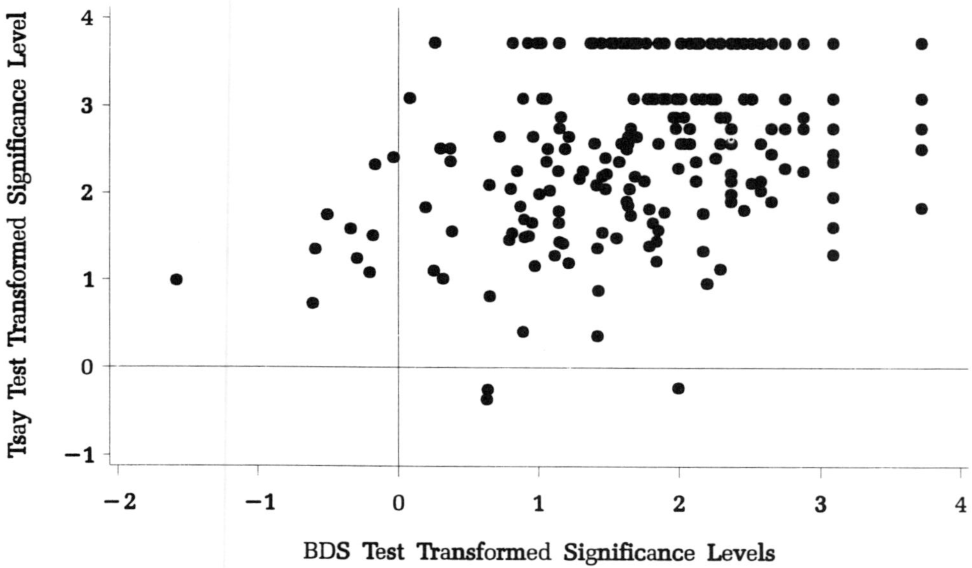

Figure 5-2 Cross-Plot of Cubic Nonmartingale Model Test Results: Transformed Significance levels for Tsay and BDS Tests

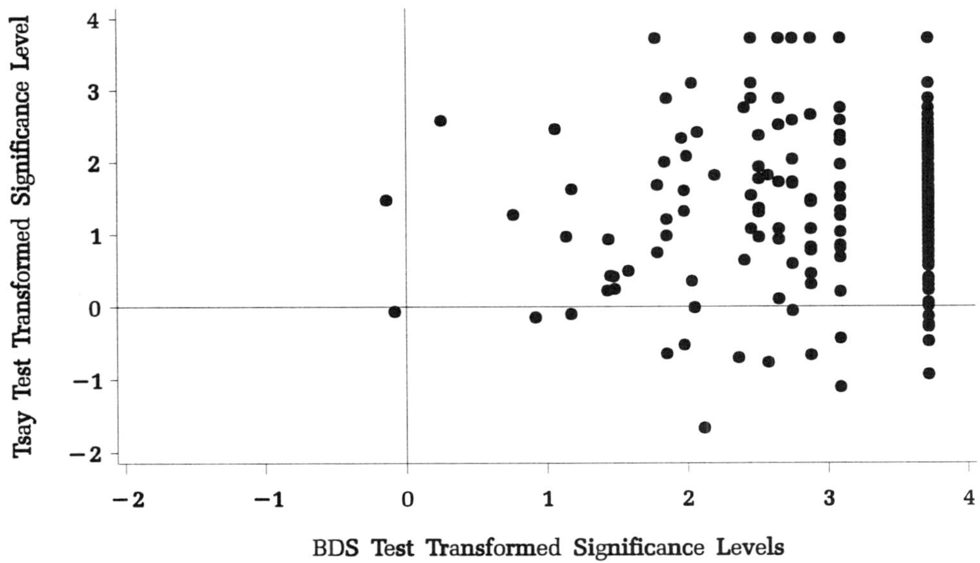

References

Ashley, R., Patterson, D. M. and Hinich, M. (1986). "A Diagnostic Test for Nonlinear Serial Dependence in Time Series Fitting Errors" *Journal of Time Series Analysis 7*, 165-78.

Barnett, W. A., A.R. Gallant, M.J. Hinich, J.A. Jungeilges, D.T. Kaplan, and M.J. Jensen (1997) "A Single-Blind Controlled Competition Among Tests for Nonlinearity and Chaos" *Journal of Econometrics 82*, 157-92.

Bollerslev, Tim (1986) "Generalized Autoregressive Conditional Heteroskedasticity" *Journal of Econometrics 31*, 307-27.

Engle, Robert F. (1982) "Autoregressive Conditional Heteroskedasticity with Estimates of the Variance of United Kingdom Inflation" *Econometrica 50*, 987-1007.

Hamilton, James (1989) "A New Approach to the Economic Analysis of Non-Stationary Time Series and the Business Cycle" *Econometrica 57*, 357-84.

Kanter, M. and Steiger W. L. (1974) "Regression and Autoregression with Infinite Variance" Advances in Applied Probability 6, 768-83.

Lam, P. (1997) "A Markov Switching Model of GNP Growth With Duration Dependence" (unpublished manuscript).

Lee, T., H. White, C.W.J. Granger (1993) "Testing for Neglected Nonlinearity in Time Series Models." *Journal of Econometrics 56*, 269-90.

Potter, S. (1995) "A Nonlinear Approach to U.S. GNP" *Journal of Applied Econometrics 10*, 109-125.

Press, W. H., Flannery, B. P., Teukolsky, S. A., and Vetterling, W. T. (1986) *Numerical Recipes: The Art of Scientific Computing*. Cambridge University Press: Cambridge.

Teräsvirta, T. and Anderson, H. (1992) "Modelling Nonlinearities in Business Cycles Using Smooth Transition Autoregressive Models." *Journal of Applied Econometrics 7*, 119-36.

Tong, H. (1983) *Threshold Models in Non-linear Time Series Analysis*, Springer-Verlag: New York.

Tong, H. and K.S. Lim (1980) "Threshold Autoregression, Limit Cycles and Cyclical Data" *Journal of the Royal Statistical Society (B) 42*, 245-92.

Notes

1.The *size* of a test – considered in Chapter 4 – is the probability that it will (wrongly) reject the null hypothesis when the null hypothesis is true; the *power* of a test is the probability that it will (correctly) reject the null hypothesis when the null hypothesis is false. Consequently, in order to examine the power of a nonlinearity test one must explicitly specify the way in which the null hypothesis is false: i.e., one must specify a particular nonlinear generating mechanism for the data as an alternative to the linear generating mechanism.

2.For example, see Ashley, et al. (1986), Lee, et al. (1993), and Barnett, et al. (1997).

3.It may be that the most relevant alternative models are the ones which have been estimated using the actual data of interest. See Chapter 9 for an example of this approach.

4.More sophisticated switching models, estimated by Potter (1995) and Lam (1997) using U.S. real GNP data, are considered in Chapter 9.

5.Normal deviates were obtained using the RAN1 and GASDEV routines given by Press, et al. (1986), respectively. The adequacy of these pseudorandom deviates for the present purpose was confirmed by comparing our results to those obtained using IMSL subroutines and also by observing that our results converge to those obtained from asymptotic theory when the sample size is large. EachStudent's t deviate was generated from its definition, using normal deviates to generate aχ^2 deviate, etc. The exact algorithm of Kanter and Steiger (1974) was used to generate symmetric stable Paretian innovations.

6.This result *is* confirmed using simulated data from the Potter (1995) TAR model for U.S. real GNP; see Table 9-2 in Chapter 9 below.

7.It is necessary to use a truncated c.d.f. because the test statistic obtained from the prewhitened simulated data in some cases exceeds that obtained in all 1000 bootstrapped samples drawn from its empirical distribution. Indeed, because of this, the transformed data are still substantially non-Gaussian, albeit far less so than the raw significance levels. Consequently, we do not present confidence intervals based on these sample correlations.

6 ANALYSIS OF STOCK MARKET RETURNS

INTRODUCTION

In this chapter we use our battery of nonlinearity tests to study the time series properties of growth rates in common stock prices at different levels of time aggregation. Financial economists call such growth rates "rates of return," and express them as decimal fractions. We also look at the effect of "hard clipping" the data in order to separate the issue of conditional dependence in variance from the deeper issue of serial nonlinear dependence in the time series. By "hard clipping" we mean a transformation that converts the data to a sequence of +1's and -1's, depending on whether the return is positive or negative.

The first comprehensive study of dependence in stock market rates of return, or simply "returns," can be attributed to Fama (1965) in a paper that is appropriately titled "The Behavior of Stock Market Prices." Fama studied the daily returns (a return measured over a single day) of the 30 stocks that made up the Dow Jones Industrial average (DJIA) at the time of his study. He found low levels of serial correlation at short lags which, when significant, were only marginally so. Fama also provided considerable evidence concerning the non-Gaussian nature of the empirical distribution of daily returns. In particular, his evidence shows that the empirical distributions of returns are too peaked and have tails which are too heavy to be consistent with the normal distribution. This characteristic of a distribution is known as "leptokurtosis." Fama's evidence of low-to-zero serial correlation led many financial economists of the time to incorrectly conclude that daily returns are serially independent, and therefore, that a random walk with drift is an appropriate statistical model of stock market behavior. (See Example 1.3 in Chapter 1.) However, zero serial correlation implies statistical independence if and only if the joint probability distribution is normal. The importance of this condition was made clear with the discovery of nonlinear dependence in stock market returns.

Nonlinearity in common stock returns was first reported by Hinich and Patterson in 1985. They studied the daily rates of return of 15 stocks selected at random from all stocks listed on the New York Stock Exchange (NYSE) and the American Stock Exchange (ASE). Strong evidence of nonlinearity was found in each of the 15 return series using Hinich's Bispectrum test. In 1986 Scheinkman and LeBaron published a paper that reported the finding of nonlinear dependence in weekly returns to the valued-weighted CRSP index.[1] These authors made use of the correlation dimension, the conditional probability that two points are close given that their past N histories are close, the Takens estimate of closeness, and the BDS statistic. Hsieh (1991) presents additional evidence on nonlinearity in stock returns based on applying the BDS test.

Bollerslev (1987) derived the Generalized Autoregressive Conditionally Heteroskeclastic (GARCH) time series model. In that paper he estimated a GARCH (1,1)-t model for the monthly Standard and Poor's 500 Composite index (S&P 500), the S&P Industrial index, the S&P Capital Goods index, and the S&P Public Utilities index. Bollerslev's results supported the presence of conditional volatility in each of these return series.[2]

For reasons that are not entirely clear, the ARCH/GARCH class of nonlinear models dominate theoretical and empirical discussions of nonlinearity in stock market returns. Bollerslev, Chow, and Kroner (1992) list over 300 papers related to the application of ARCH/GARCH models in finance applications. Perhaps their popularity is related to the martingale difference property of ARCH/GARCH models, which rules out the ability to make point forecasts of security returns. This fact makes ARCH/GARCH consistent with the weak form of the so-called "Efficient Markets Hypothesis" of financial economics. Or their popularity could flow from the fact that ARCH/GARCH offers specific models that can be fitted to data. Hsieh (1995) attributes the popularity of ARCH/GARCH to the ability of a fitted model to provide "an estimate of daily volatility, which can be used to standardize the return series." He goes on to say, "The resulting series typically exhibits little remaining nonlinearity." The problem with the latter statement is that it ignores the fact that ARCH/GARCH is a model and not a test, and that an ARCH/GARCH model is easily misspecified. For example, GARCH (1,1) models can be fit, with highly significant parameter estimates, to nonlinear quadratic models similar to those shown in Table 5-1.[3] However, the ability of an ARCH/GARCH model to remove most of the nonlinearity in a time series cannot be denied. In this role the ARCH-type models can be regarded as a generic nonlinear deconvolver. A second problem with the view that the only type of nonlinearity present in stock returns is ARCH/GARCH is the inherent improbability of the claim. Given the uncountable number of possible nonlinear stochastic processes that might be generating stock returns, it is unlikely that only a tiny subset – those that are both a martingale difference and that involve only fourth-order moments – can be found generating stock market returns. As we will see in the remaining sections of this chapter there is considerable evidence to support the presence of other forms of nonlinearity in stock returns.

The evidence of nonlinearity in stock returns is not limited to the study of daily, weekly, or monthly returns. Hinich and Patterson (1989) analyze 15-minute returns for 15 of the DJIA stocks over a three-year period. They found, using the Hinich Bispectrum test, that linearity was rejected at extremely small significance levels for all 15 stocks. In a later paper, Hinich and Patterson (1993) used the 15-minute returns of their 1989 paper to examine whether or not the incidence of nonlinearity was a function of the day of the week, or the time of day. They found no evidence of a day-of-the-week effect, but the analysis suggests that the incidence of nonlinearity is most frequent during the first and last 30 minutes of the trading day. Andersen and Bollerslev (1997) argue that intra day volatility shocks caused by short-run information arrivals may be responsible for long-run dependence in volatility.

Theoretical models of the continuous time stochastic process generating stock market prices are usually based on a diffusion process as we discussed in Chapter 1.

Diffusion processes have continuous sample paths. In reality, stock market prices are some form of a discrete point process. Recently, Engle and Lunde (1998), and Russell and Engle (1998), have developed and estimated models of the continuous time stock price process using marked point processes. Engle and Lunde find that prices respond more quickly when new information is flowing into the market, so that the price impact of trades and the volatility of prices are high in such circumstances.

Those readers who would like to learn more about the statistical properties of stock market prices are directed to *The Econometrics of Financial Markets* by Campbell, Lo, and Mackinlay (1997).

DAILY RETURNS

In this section we study the behavior of the S&P 500 index. We consider both serial correlation and nonlinearity in the daily rates of return to this index. The S&P 500 index is an aggregate measure of U.S. stock market performance. Its component securities are the 500 largest corporations traded on the U.S. capital market as measured by their capitalized market value. The S&P 500 is considered to be one of the key measures of stock market performance. The source of the S&P 500 returns used here is the CRSP data base. As of this writing CRSP provides daily S&P 500 returns starting on July 3, 1962 and ending on December 31, 1997, for a total of 8938 days. All of this data is included in our analysis.

The behavior of an aggregate stock market measure can differ in certain ways from its component securities. A phenomenon known as "nonsynchronous trading" may cause an index of stock returns to exhibit substantial autocorrelation even though the component returns have zero autocorrelation.[4] The reason for this is that the last trade of the day occurs at different times for different stocks, and that daily returns are measured from the close on day t to the close on day t + 1. The greater the variance of end of day trade times, the greater the correlation. To the extent that nonsynchronous trading causes autocorrelation, this linear dependence is an artifact of the data sampling process and has no economic meaning. This effect induces positive serial correlation primarily at lag 1, but it can also induce positive correlation at longer lags if there is a tendency for stocks to go for days at a time without trading. Nonsynchronous trading can also induce autocorrelation into weekly and monthly index returns.

We carry out the analysis by first dividing the data into 4 groups of 2000 observations each and a final group of 938 observations. Some descriptive statistics of each data group is shown below in Table 6-1. The surprise result is the kurtosis statistic, which for reference purposes is 3.0 for a normal probability distribution. Kurtosis greater than 3.0 indicates leptokurtosis: heavy or "fat" tails. As can be seen, the periods of 7/22/70 - 6/21/78 and 6/22/78 - 5/20/86 exhibit sample kurtosis less than 3.0, which is called "platykurtosis," and which implies a smaller tail-frequency and a flatter top than the normal distribution. The kurtosis statistic, in contrast, is extremely high for the period 5/21/86 - 4/15/94. Figure 6-1 is plot of the returns for this particular period. Horizontal lines are drawn at ± one sample standard deviation.

The extreme gyrations in S&P 500 returns around late October 1987 are a reflection of the famous '87 stock market crash: on October 19, 1987 ("Black Monday") the S&P 500 index fell by more than 20%. Clearly, this event must have an impact on the sample kurtosis for the period 5/21/86 - 4/15/94 shown in Table 6-1. In order to demonstrate the impact, we trimmed back the return on "Black Monday" to minus two sample standard deviations (-0.022) and recalculated the sample kurtosis. The new kurtosis statistic is 10.512, still high with respect to a normal distribution, but not out of line with the kurtosis for the 4/18/94 - 12/31/97 period.

Table 6-1 Some Descriptive Statistics of Daily S&P 500 Returns

Period	Mean	Std. Dev.	Kurtosis
7/3/62 - 7/21/70	0.00018	0.006	4.979
7/22/70 - 6/21/78	0.00015	0.009	2.031
6/22/78 - 5/20/86	0.00049	0.009	1.600
5/21/86 - 4/15/94	0.00038	0.011	73.017
4/18/94 - 12/31/97	0.00118	0.008	8.481

Table 6-2 shows the autocorrelation function for lags of 1 - 5 days for each of the five sub periods. There is significant lag 1 positive correlation, an indication of nonsynchronous trading, during the first three sub periods. The dramatic increase in the volume of stock market trading since the mid-1980s was probably accompanied by less heterogeneity in end-of-day trade times, which would lower the impact of nonsynchronous trading on autocorrelation. This could also explain the lack of significant lag 1 correlation for the last two sub periods.

In Table 6-3 we display significance levels for the nonlinearity tests. All the tests were done with 1000 bootstraps. The McLeod-Li test and the Engle test are calculated for 8 lags and 5 lags, respectively. The BDS test is reported for $\varepsilon = 1.0$, as in the earlier chapters. The Tsay and Bicovariance tests are computed with 9 lags, and the Bispectrum test was run with a smoothing window width of 17. The Table shows that, with few exceptions, every test rejects all the time, so in a sense we haven't learned much we didn't already know: the sample sizes in Table 6-3 are on the large side, and we are reminded of the old admonition that given enough data you can reject almost anything. With this in mind, we next consider the year-by-year records.

A CLOSER LOOK AT DAILY RETURNS

Table 6-4 shows the serial correlation in the daily S&P 500 returns on a year-by-year basis from the second half of 1962 through 1997. We see strong serial correlation at

Figure 6-1 Plot of S & P 500 Daily Returns for the Period 5/21/86 - 4/15/94

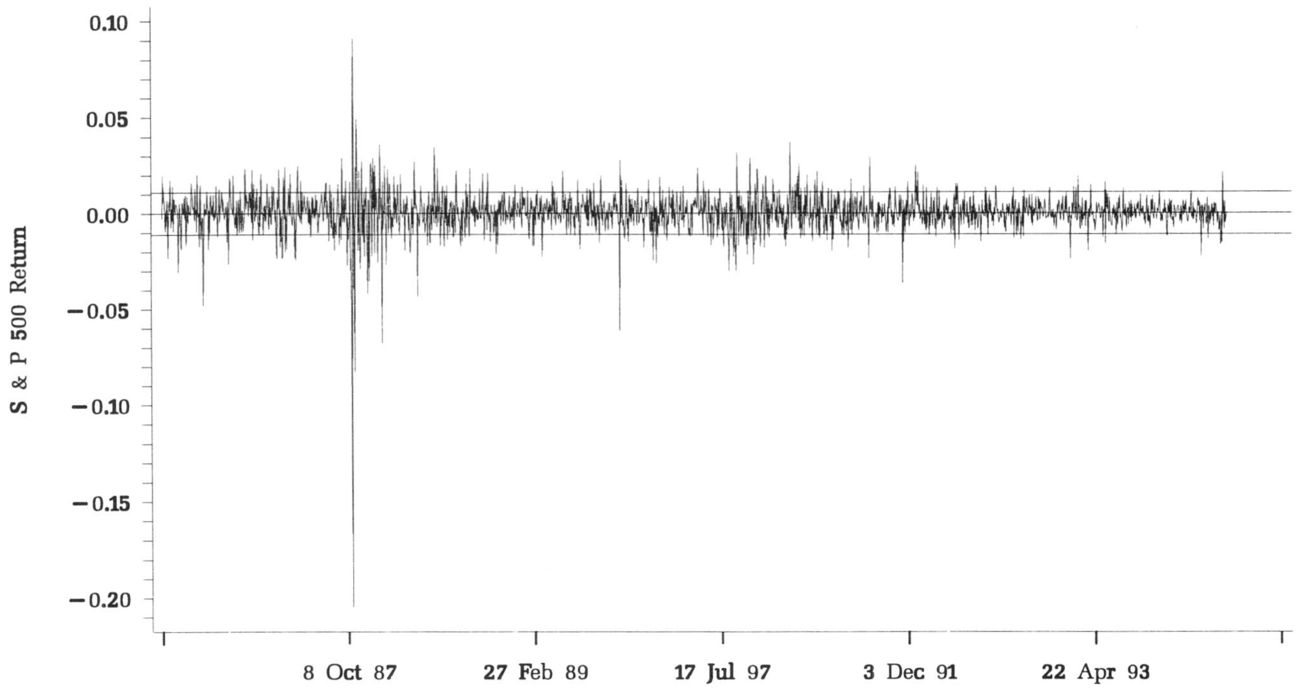

Table 6-2 Serial Correlation in Daily S&P 500 Growth Rate

Lag	Time Period[1]				
	7/3/62 - 7/21/70	7/22/70 - 6/21/78	6/22/78 - 5/20/86	5/21/86 - 4/15/94	4/18/94 - 12/31/97
1 day	0.219*	0.253*	0.093*	0.030	0.020
2 days	0.011	-0.016	0.028	-0.074*	-0.047
3 days	0.042	0.018	-0.021	-0.043	-0.055
4 days	0.049*	-0.005	-0.026	-0.050*	-0.051
5 days	0.039	-0.045	0.007	0.065*	-0.039

[1] All periods contain 2000 observations except 4/18/94 - 12/31/97 which contain 938 observations.
* Indicates significant at 5% level.

Table 6-3 Significance Levels for Nonlinearity Tests on Daily S&P 500 Growth Rate[1]

Test	7/3/62 - 7/21/70	7/22/70 - 6/21/78	6/22/78 - 5/20/86	5/21/86 - 4/15/94	4/18/94 - 12/31/97
McLeod - Li	0.000	0.000	0.000	0.010	0.004
Engle LM	0.000	0.000	0.000	0.003	0.003
BDS					
m = 2	0.000	0.000	0.097	0.000	0.000
m = 3	0.000	0.000	0.045	0.000	0.000
m = 4	0.000	0.000	0.011	0.000	0.000
Tsay	0.000	0.000	0.000	0.000	0.000
Bicovariance	0.000	0.000	0.000	0.000	0.000
Bispectral	0.311	0.011	0.000	0.727	0.163

[1] All time periods 2000 observations except 4/18/94 - 12/31/97 which contain 938 observations.

Table 6-4 Serial Correlation of S&P 500 Daily Growth Rate: The Year-by-Year Results[1]

	Year								
Lag	1962	1963	1964	1965	1966	1967	1968	1969	1970
1 day	0.037	-0.058	0.135*	0.193*	0.259*	0.139*	0.234*	0.337*	0.362*
2 days	0.058	-0.033	0.054	-0.036	0.020	-0.033	0.030	-0.006	0.012
3 days	0.185*	0.052	-0.010	0.041	-0.074	0.101	0.160*	-0.012	0.025
4 days	0.008	-0.041	0.040	-0.012	-0.001	0.188*	0.110	0.075	0.063
5 days	0.007	0.010	-0.023	0.025	0.043	0.058	0.092	-0.000	0.049

[1] Sample size is nominally 250 observations for each year with the exception of 1962, which contains 125 observations.

* Significant at the 5% level.

Table 6-4 (continued) Serial Correlation of S&P 500 Daily Growth Rate: The Year-by-Year Results

	Year								
Lag	1971	1972	1973	1974	1975	1976	1977	1978	1979
1 day	0.249*	0.276*	0.237*	0.254*	0.232*	0.146*	0.206*	0.127*	0.162*
2 days	0.016	0.027	-0.067	-0.004	-0.010	-0.040	0.034	0.092	-0.054
3 days	0.037	0.026	-0.127	0.058	0.008	0.048	-0.064	0.067	-0.020
4 days	0.053	-0.104	-0.077	-0.031	0.008	-0.045	-0.012	0.022	0.041
5 days	0.035	-0.024	-0.101	-0.148	0.023	-0.030	-0.010	0.035	0.013

Table 6-4 (continued) Serial Correlation of S&P 500 Daily Growth Rate: The Year-by-Year Results

	Year								
Lag	1980	1981	1982	1983	1984	1985	1986	1987	1988
1 day	0.1134	0.099	0.110	0.004	0.114	0.068	0.103	0.039	-0.032
2 days	0.033	-0.037	0.055	0.055	0.028	-0.056	-0.044	-0.140*	-0.060
3 days	-0.037	-0.053	0.080	-0.119	-0.045	-0.016	-0.008	-0.054	-0.058
4 days	-0.132	-0.065	0.030	-0.023	-0.052	0.016	0.047	-0.080	-0.018
5 days	0.105	-0.010	-0.042	-0.106	0.086	0.040	-0.035	0.148*	-0.058

Table 6-4 (continued) Serial Correlation of S&P 500 Daily Growth Rate: The Year-by-Year Results

	Year								
Lag	1989	1990	1991	1992	1993	1994	1995	1996	1997
1 day	-0.071	0.110	0.036	-0.039	-0.008	-0.004	-0.002	0.144	-0.040
2 days	0.028	-0.062	0.049	-0.043	-0.005	0.111	-0.096	0.041	-0.100
3 days	-0.013	-0.069	-0.013	-0.056	-0.043	-0.077	-0.107	-0.064	-0.031
4 days	-0.174	0.000	0.063	-0.022	-0.089	-0.088	0.004	-0.040	-0.043
5 days	0.003	0.034	0.034	-0.023	-0.025	-0.055	-0.096	-0.065	-0.029

lag 1 from 1964 through 1979, which roughly corresponds to what we saw in Table 6-2. Recent research has questioned exactly how much serial correlation nonsynchronous trading is capable of explaining. Lo and MacKinlay (1990) conclude that nonsynchronous trading is capable of explaining only a small portion of the autocorrelation that is observed in the data. Boudoukh, Richardson, and Whitelaw (1994) show that inferences from nonsynchronous trading models are highly sensitive to assumptions regarding nontrading intervals and other stock market parameters, and argue that prior studies *understate* the effects of nonsynchronous trading. Kadlec and Patterson (1999) find that simulated indices (portfolios) calculated using actual end-of-day trade times for a sample of 400 stocks generate daily returns which exhibit levels of autocorrelation that are about one-half that of the observed daily returns for the portfolios. Further, they find that indices of simulated weekly portfolios exhibit levels of autocorrelation that are roughly one-fourth that of their observed weekly returns.

The year-by-year results of the linearity tests are displayed in Table 6-5. For the analysis of the yearly returns we set the test parameters as follows: McLeod-Li to 8 lags, Engle test to 5 lags, ε for BDS remained at 1.0, Bicovariance test at 9 lags and Tsay test at 6 lags, the smoothing window for the Bispectrum was set to 17. We again used 1000 bootstraps. In looking over this rather lengthy table we see that the incidence of nonlinearity varies considerably by year. There is a tendency to reject linearity more frequently before the mid-1980s than after. Surprisingly, we find a number of years where the BDS test fails to reject at the 5% level, yet one or more of the other tests does reject at 5%; this is the case for about one-third of the years shown.

In the years 1976, 1979, 1981, 1984, 1989, 1994, and 1996 neither the McLeod-Li nor the Engle tests (which should be sensitive to ARCH-type behavior) reject at the 5% level or less, whereas at least one of the third-order cumulant sensitive tests does reject at 5% or less. With the exception of 1996, the BDS test also fails to reject in these same years. We conclude that in these years the form of the nonlinearity present in the returns is somehow much different than in other years. Given the nature of the tests involved it appears that ARCH-type effects are greatly attenuated, or not present at all, in these years.

The tests for 1987 were rerun with the trimmed return for October 19. The only substantive change from the results in Table 6-5 is that the p-value for the Bispectral test is now 0.041, compared with 0.093 in the Table.

The ARCH-type models are examples of the martingale difference class of stochastic processes; see the section "Martingale Differences" in Chapter 1. If we hard clip a time series generated by a symmetric ARCH-type model we will remove this form of dependence in the data and be left with an iid series. By "hard clip" we mean the following binary transformation of the data $\{x_t\}$ to $\{y_t: y_t = 1 \text{ if } x_t \geq 0 \text{ and } y_t = -1 \text{ if } x_t < 0\}$. Thus if $\{x_t\}$ is an ARCH or GARCH process with a symmetric distribution, the sequence $\{y_t\}$ is a binary pure noise process.[5] On the other hand, if the transformed series shows some form of dependence, then we may conclude that the source series $\{x_t\}$ was not a symmetric ARCH or GARCH series.

This binary transformation was applied to the 1967 daily stock returns. We chose 1967 because it is a year where all of the nonlinearity tests are highly significant.

Table 6-5 Significance Levels for Nonlinearity Tests on Daily S&P 500 Growth Rate: The Year-by-Year Results[1]

Test	1962	1963	1964	1965	1966	1967	1968	1969	1970
McLeod - Li	0.000	0.004	0.024	0.000	0.000	0.000	0.036	0.004	0.015
Engle LM	0.001	0.009	0.019	0.000	0.001	0.000	0.047	0.001	0.023
BDS									
m = 2	0.000	0.021	0.000	0.000	0.032	0.000	0.393	0.044	0.122
m = 3	0.000	0.000	0.000	0.000	0.000	0.000	0.041	0.000	0.010
m = 4	0.000	0.000	0.000	0.000	0.000	0.000	0.006	0.000	0.008
Tsay	0.137	0.000	0.011	0.000	0.001	0.015	0.396	0.164	0.001
Bicovariance	0.000	0.000	0.032	0.000	0.001	0.001	0.088	0.499	0.002
Bispectral	0.342	0.174	0.083	0.119	0.039	0.042	0.199	0.654	0.075

Sample sizes are nominally 250, with the exception of 1962 which is 125 observations.

Table 6-5 (continued) Significance Levels for Nonlinearity Tests on Daily S&P 500 Growth Rate: The Year-by-Year Results

Test	1971	1972	1973	1974	1975	1976	1977	1978	1979
McLeod - Li	0.036	0.001	0.001	0.010	0.533	0.226	0.960	0.025	0.093
Engle LM	0.019	0.047	0.002	0.012	0.762	0.134	0.981	0.022	0.672
BDS									
m = 2	0.031	0.024	0.235	0.383	0.627	0.225	0.284	0.003	0.371
m = 3	0.034	0.173	0.031	0.332	0.507	0.249	0.234	0.004	0.477
m = 4	0.055	0.404	0.041	0.207	0.458	0.296	0.278	0.005	0.597
Tsay	0.014	0.149	0.295	0.023	0.277	0.165	0.521	0.020	0.347
Bicovariance	0.006	0.053	0.065	0.072	0.016	0.703	0.436	0.002	0.494
Bispectral	0.297	0.004	0.005	0.014	0.458	0.036	0.839	0.019	0.001

Table 6-5 (continued) Significance Levels for Nonlinearity Tests on Daily S&P 500 Growth Rate: The Year-by-Year Results

Test	1980	1981	1982	1983	1984	1985	1986	1987	1988
McLeod - Li	0.047	0.293	0.014	0.127	0.235	0.391	0.897	0.040	0.052
Engle LM	0.034	0.476	0.021	0.185	0.144	0.477	0.971	0.033	0.650
BDS									
m = 2	0.575	0.855	0.137	0.543	0.367	0.915	0.778	0.000	0.314
m = 3	0.619	0.908	0.039	0.754	0.446	0.849	0.851	0.000	0.412
m = 4	0.560	0.650	0.008	0.674	0.442	0.799	0.859	0.000	0.395
Tsay	0.824	0.597	0.348	0.503	0.217	0.115	0.705	0.001	0.389
Bicovariance	0.046	0.794	0.027	0.077	0.034	0.152	0.536	0.000	0.783
Bispectral	0.611	0.021	0.045	0.281	0.075	0.158	0.309	0.093	0.269

Table 6-5 (continued) Significance Levels for Nonlinearity Tests on Daily S&P 500 Growth Rate: The Year-by-Year Results

Test	1989	1990	1991	1992	1993	1994	1995	1996	1997
McLeod - Li	0.278	0.001	0.508	0.008	0.234	0.150	0.329	0.074	0.031
Engle LM	0.119	0.007	0.260	0.017	0.230	0.200	0.393	0.892	0.020
BDS									
m = 2	0.773	0.011	0.824	0.149	0.238	0.193	0.651	0.047	0.643
m = 3	0.582	0.021	0.400	0.075	0.477	0.113	0.615	0.015	0.265
m = 4	0.397	0.029	0.156	0.075	0.569	0.055	0.515	0.008	0.198
Tsay	0.248	0.109	0.209	0.092	0.605	0.171	0.330	0.010	0.006
Bicovariance	0.044	0.000	0.605	0.131	0.439	0.025	0.072	0.014	0.011
Bispectral	0.120	0.008	0.759	0.050	0.089	0.306	0.696	0.752	0.571

For the 1967 daily data, a prewhitening model was not needed under the Schwartz criterion. However, the binary data required an AR(1) model ($\varphi_1 = 0.167$) for prewhitening, which indicates the presence of other forms of serial dependence in the data beyond ARCH or GARCH. Below in Table 6-6 we display the significance levels for the nonlinearity tests run on the binary data. Note that those tests which should be particularly sensitive to ARCH/GARCH effects – i.e., the McLeod-Li and Engle tests – are not rejecting the null hypothesis of serial independence at anything close to conventional significance levels. The BDS test, which is sensitive to many forms of nonlinearity, is significant at 5% level for m = 2 and m = 3. Coupled with the evidence that the binary series now exhibits serial dependence at lag 1, we conclude that ARCH-type models are not solely responsible for our rejection of linearity in the original series.

Table 6-6 Significance Levels for 1967 Binary Data

McLeod-Li	0.930
Engle LM	0.984
BDS	
m = 2	0.031
m = 3	0.050
m = 4	0.062
Tsay	0.832
Bicovariance	0.815
Bispectral	0.883

Hsieh (1991) examines S&P 500 daily returns as well as weekly returns for the seven year period of 1983 - 1989. Using a third-order moment test which is a modification of the Tsay test he finds no evidence to reject the null hypothesis that stock returns have zero third-order moments.[6] Looking at years 1983 - 1987, we see little evidence of any form of nonlinearity, so in this regard the Hsieh result is not surprising. Hsieh, however, rejects linearity using the BDS test, which is inconsistent with Table 6-5 except for 1987. It is possible, given the high power of the BDS test, that 1987 could cause the BDS test to reject over the entire 1983 - 1989 sample; it is more likely that Hsieh's BDS result is due to the fact that he did not bootstrap the test.

Brock, Lakonishok, and LeBaron (1992) take a different approach to identifying the presence of GARCH in mean and exponential GARCH (EGARCH) in the daily levels of the Dow Jones Industrial Average. First, they applied two "technical" trading rules to the data; these rules produce buy and sell signals for an investor based on past price history. They noted the frequency of buy (sell) signals as well as the profits obtained. Next, they simulated the Dow Jones Industrial Average using four different fitted models: Random Walk, AR(1), GARCH in mean, and EGARCH. The trading rules were then run against the simulated (bootstrapped) series. The authors found that the technical trading rules produced patterns of buy and sell signals (and profits) that were unlike those generated by the actual Dow Average. They concluded that "The returns obtained from buy (sell) signals are not likely to be generated by the four popular null models."

WEEKLY RETURNS

Daily returns were converted to weekly returns by summing the daily returns over a calendar week: Friday close to Friday close. Because of holidays, some weeks contain only 4 trading days. The period January 2, 1963 through December 21, 1997 yielded 1820 weekly returns. The data set was then divided into 5 year subsets of 260 observations each. The sample size of 260 is very close to the nominal size of 250 in the year-by-year daily returns (sample sizes varied from 250 to 254 per year). As a consequence, differences in the behavior of weekly versus daily returns will not be attributable to respective sample sizes, but rather to the time aggregation of the returns.

We start the analysis by presenting the key sample statistics of the weekly returns in Table 6-7. Because each return is measured over a longer time period we

Table 6-7 Descriptive Statistics of Weekly S&P Returns

Time Period	Mean	Std. Dev.	Kurtosis
1963 - 67	0.0017	0.013	2.396
1968 - 72	0.009	0.017	1.022
1973 - 77	-0.0006	0.025	3.344
1978 - 82	0.0017	0.022	1.211
1983 - 87	0.0025	0.022	3.618
1988 - 92	0.0024	0.019	0.690
1993 - 97	0.0031	0.015	0.573

would expect the mean and standard deviations shown in Table 6-7 to be larger than was the case for daily returns; this is true, with the exception of the mean for 1973 - 1977. The important fact to be gleaned from Table 6-7 is that the kurtosis statistics for weekly returns are much lower than for daily returns.

Table 6-8 displays the sample autocorrelation function for the weekly S&P 500 returns (growth rates). The 1968 - 1972 period is the only time period where we see statistically significant positive correlation at lag 1. The negative correlation at lag

1 for 1973 - 1997 is significant in the statistical sense, but cannot be explained by nonsynchronous trading: nonsynchronous trading can only explain positive correlation. Therefore, some other form of linear dependence is at work during those time periods.

The results of the linearity tests are shown in Table 6-9. We see that evidence supporting the nonlinear behavior of weekly S&P 500 returns is strong. The parameter values for each test were set to the same values used with the daily returns. For daily returns, as shown in Table 6-5, the incidence of nonlinearity post 1980 is more episodic than is the case for weekly S&P 500 returns. In particular, the BDS test rejects at 5% or lower for every period shown in Table 6-9. By comparison, the BDS test rejects at 5% or lower for approximately 2/3 of the years studied with daily returns. We don't believe that the difference between daily and weekly returns represents some sort of time aggregation bias with respect to nonlinear dependence. The difference between the results of the daily and weekly returns more likely reflect the fact that the weekly sample spans a longer total time interval in which there is a greater chance for a nonlinear episode to occur.

MONTHLY S&P 500 RETURNS

Monthly returns were created by time aggregating weekly S&P 500 index returns. This process yielded 426 monthly returns from mid 1962 through the end of 1997. These were divided into two sub periods: July 1962 - December 1980, and January 1981 - December 1997. The sample sizes are 222 and 204, respectively, which is approximately the same size as the daily and weekly data sets.

In Table 6-10 we show the usual descriptive statistics for the monthly S&P 500 returns. As we would expect, both the sample means and the sample standard deviations are higher than was the case for the weekly S&P 500 index returns. The kurtosis statistic is roughly similar to its values for the weekly data.

The sample autocorrelation function for monthly returns is consistent with a white noise process for each sample period. The Box-Ljung portmanteau test for lags 1 - 6 is only significant at p-values of 0.253 and 0.970 for the first and second sample periods, respectively.

Nonlinearity is a form of stochastic dependence. It makes sense that the intensity of this dependence declines with increases in the time between observations. Think of this as a loss of resolution as the measurement interval becomes longer – the nonlinear stochastic signal is transient in nature, and therefore will not be observable if the time between observations is sufficiently long. That is, the signal we are trying to detect will come and go between distant observations. Hence the question: At what sampling interval will we no longer be able to detect nonlinear dependence? Christoffersen and Diebold (1997) report that the conditional volatility in the daily S&P 500 index – the behavior modeled by ARCH and GARCH – decays to an insignificant level after ten trading days. Their data ran from January 1973 through mid 1977. We should expect to see, therefore, little evidence of nonlinearity in those tests that are sensitive to ARCH and GARCH in the monthly S&P 500 index returns, at least for the time period studied by Christoffersen and Diebold.

Table 6-8 Serial Correlation of Weekly S&P 500 Growth Rate[1]

	Time Period						
Lag	1963 - 67	1968 - 72	1973 - 77	1978 - 82	1983 - 87	1988 - 92	1993 - 97
1 day	0.100	0.169*	-0.128*	0.061	0.068	-0.168*	-0.150*
2 days	0.064	-0.024	-0.051	-0.002	0.002	0.102	0.097
3 days	-0.214	-0.003	0.279*	-0.004	-0.045	-0.016	-0.023
4 days	0.031	-0.057	-0.116	0.022	0.007	0.000	-0.097
5 days	-0.079	0.056	-0.103	-0.035	-0.031	-0.054	-0.121

[1] Sample size is 260 observations for each 5 year period.

Table 6-9 Significance Levels for Nonlinearity Tests on Weekly S&P 500 Growth Rate[1]

Test	1963 - 67	1968 - 72	1973 - 77	1978 - 82	1983 - 87	1988 - 92	1993 - 97
McLeod - Li	0.000	0.000	0.000	0.063	0.000	0.044	0.000
Engle LM	0.000	0.001	0.000	0.172	0.004	0.027	0.007
BDS							
m = 2	0.000	0.038	0.002	0.017	0.010	0.004	0.156
m = 3	0.000	0.026	0.003	0.018	0.036	0.004	0.037
m = 4	0.000	0.016	0.002	0.011	0.052	0.003	0.004
Tsay	0.000	0.021	0.000	0.014	0.022	0.524	0.011
Bicovariance	0.000	0.000	0.000	0.130	0.000	0.592	0.005
Bispectral	0.017	0.007	0.006	0.282	0.060	0.750	0.466

[1] The sample size is 260 observations for each period shown.

Table 6-10 Descriptive Statistics of Monthly S&P 500 Returns

Time Period	Mean	Std. Dev.	Kurtosis
July 62 - Dec. 80	0.0042	0.041	1.412
Jan. 81 - Dec. 97	0.010	0.040	3.373

Table 6-11 Significance Level for Nonlinearity Tests on S&P 500 Monthly Growth Rate

Test	July 1962 - Dec. 1980[2]	Jan. 1981- Dec. 1997[2]
McLeod-Li	0.141	0.957
Engle LM	0.071	0.980
BDS		
m = 2	0.215	0.785
m = 3	0.177	0.869
m = 4	0.022	0.664
Tsay	0.070	0.571
Bicovariance	0.133	0.943
Bispectral	0.014	0.794

Table 6-11 displays the results of running the battery of nonlinearity tests on the monthly return data. For this table the McLeod-Li test uses 8 lags and the Engle test 5 lags. The Bicovariance and Tsay tests are calculated using 8 lags. The Hinich Bispectrum test uses a smoothing window of width 16. The Table shows no evidence of nonlinearity in the second sample period, January 1981 - December 1997. In the first period, July 1962 - December 1980, only the Bispectrum test, and the BDS test at embedding dimension 4, have p-values less than 0.05 (that is, they are significant at the 5% level). The Engle test and the Tsay test are significant at the 10% level. The contrast with the weekly returns is dramatic, where we did see much stronger evidence of nonlinearity.

---

1. Number observations is 222.
2. Number observations is 204.

We conclude that the strong nonlinear dependence exhibited at daily and weekly sample intervals is greatly attenuated at the monthly interval. The lack of either linear or nonlinear dependence in monthly returns since the early 1980s suggests that the geometric random walk with drift model presented in Chapter 1 is a good description of stock market behavior at this level of time aggregation.

Notes

1. CRSP is an acronym for Center for Research in Security Prices which is located at the Graduate School of Business, University of Chicago. It's valued weighted index contains every stock on the NYSE and ASE weighted by the capitalized market value of each stock expressed as a proportion of the total market value of all stocks in the index.

2. See Nelson (1991) and Hsieh (1995) for additional evidence on the presence of ARCH/GARCH effects in stock market returns.

3. We don't mean to imply that the available linearity tests are capable of reliably discriminating between ARCH-type processes and other processes; at least not in the finite samples. Rather, we are saying that a fitted model is a poor substitute for identification techniques.

4. For details on nonsynchronous trading see Fisher (1966) and Scholes and Williams (1977).

5. See Hinich and Patterson (1998) for additional details on this binary transformation. Note that EGARCH is conditionally skewed and therefore not symmetric.

6. The null in the bicovariance and bispectral tests is statistical independence. ARCH/GARCH processes are included under the null of the Hsieh test.

REFERENCES

Andersen, T. and T. Bollerslev (1997) "Heterogeneous Information Arrivals and Return Volatility Dynamics: Uncovering the Long-Run in High Frequency Returns,"*Journal of Finance* 52, 975-1005.

Bollerslev, T. (1987) "A Conditionally Heteroskedastic Time Series Model for Speculative Prices and Rates of Return," *The Review of Economics and Statistics* 69, 542-547.

______, R. Chow, and K. Kroner (1992) "ARCH Modeling in Finance: A Selective Review of the Theory and Empirical Evidence," *Journal of Econometrics* 52 (1-2), 5-59.

Boudoukh, J., M. Richardson and R. Whitelaw (1994) "A Tale of Three Schools: Insights on Autocorrelations of Short-Horizon Returns," *Review of Financial Studies* 7, 539-573.

Brock, W., J. Lakonishok and B. LeBaron (1992) "Simple Technical Trading Rules and the Stochastic Properties of Stock Returns," *Journal of Finance* 47, 1731-1764.

Campbell, J., A. Lo, and A. Mackinlay (1997)*The Econometrics of Financial Markets*, Princeton University Press, Princeton.

Christoffersen, P. and F. Diebold (1997) "How Relevant is Volatility Forecasting for Financial Risk Management," Wharton Financial Institutions Center Working Paper 97-45.

Engle, R. and A. Lunde (1998) "Trades and Quotes: A Bivariate Point Process," Working Paper, Dept. of Economics, University of California – San Diego.

Fama, E. (1965) "The Behavior of Stock Market Prices," *Journal of Business* 38, 34-105.

Fisher, L. (1966) "Some New Stock Market Indices," *Journal of Business* 39, 191-225.

Hinich, M. and D. Patterson (1985) "Evidence of Nonlinearity in Daily Stock Returns,"*Journal of Business and Economic Statistics* 3, 69-77.

______ (1989) "Evidence of Nonlinearity in the Trade-by-Trade Stock Market Generating Process," in W. Barnett, J. Geweke, and K. Shell, eds., *Economic Complexity: Chaos, Sunspots, Bubbles and Nonlinearity*, Cambridge University Press, Cambridge, MA, 383-409.

______ (1993) "Intra day Nonlinear Behavior of Stock Prices," in R. Day and P. Chen eds.*Nonlinear Dynamics and Evolutionary Economics*, Oxford University Press, New York, 201-214.

______ (1998) "The Episodic Behavior of Dependencies in High Frequency Stock Returns," Working Paper, Dept. of Finance, Virginia Polytechnic Institute and State University.

Hsieh, D. (1991) "Chaos and Nonlinear Dynamics: Application to Financial Markets,"*Journal of Finance* 46, 1839-1877.

______ (1995) "Nonlinear Dynamics in Financial Markets: Evidence of Implications,"*Financial Analysts Journal* July-August, 55-62.

Kadlec, G. and D. Patterson (1999), "A Transactions Data Analysis of Nonsynchronous Trading,"*Review of Financial Studies*, forthcoming.

Lo, A. And A. Mackinlay (1990) "An Econometric Analysis of Nonsynchronous Trading,"*Journal of Econometrics* 45, 181-211.

Nelson, D. (1991) "Conditional Heteroskedasticity in Asset Returns: A new Approach,"*Econometrica* 59, 347-370.

Russell, J. and R. Engle (1998) "Econometric Analysis of Discrete-Valued Irregularly Spaced Financial Transactions Data Using a New Autoregressive Conditional Multinomial Model," Working Paper, Graduate School of Business University of Chicago.

Sckeinkman, J. and B. LeBaron (1989), "Nonlinear Dynamics and Stock Returns,"*Journal of Business* 62, 311-337.

Scholes, M. and J. Williams (1977) "Estimating Betas from Nonsynchronous Data,"*Journal of Financial Economics* 5, 309-327.

7 GLINT TRACKING ERRORS IN RADAR[1]

INTRODUCTION

Angle glint or "glint noise" refers to fluctuations in the measured angle of arrival of backscattered electromagnetic waves transmitted from a radar device. Active sonar tracking systems are also subject to glint noise for the same physical reason: When a target consists of many point reflectors, the backscattered waves from the individual elements interfere with one another to produce fluctuations in the echo signal arriving at the receiver. As the aspect angles of the individual elements change, the relative amplitudes and phases of the scattered components also change, leading to a scintillation in the amplitude of the return signal and glint noise components in the measurement of angle. These tracking errors are most pronounced at short and intermediate target ranges: at long range the target appears to be a single point within the resolution cell of the tracking system. Glint induced measurement errors can result in a situation where the tracking point lies entirely outside the physical confines of the target.

Work on methods to mitigate the magnitude of the amplitude scintillation and the variance of angle glint is an important problem in radar/sonar engineering. Solutions have important military and civilian applications. This is part of the larger problem of statistical target classification using active tracking systems.

Traditionally, the most popular method for dealing with tracking noise in active systems – whether they be radar or sonar – has been filtering. The matched filter and the deconvolver are two prime examples. Their effectiveness – and no one disputes that they are effective – is a result of their ability to remove out-of-band noise, and the fact that glint tracking errors tend to be highly autocorrelated. However, the matched-filter and the deconvolver are both linear filters: consequently, to the extent that glint noise contains nonlinear serial dependencies, the performance of the linear filter methods will be less than predicted.

In the next section we present a theoretical model of glint noise which is a nonlinear stochastic process. Later, we will study the glint errors from a radar that is tracking an aircraft target. We will look at the estimated power spectrum of the glint errors as well as the results of our battery of nonlinearity tests.

STATISTICAL DEPENDENCE IN TRACKING ERRORS

The basic physics of glint noise are well understood, the earliest models having been developed in the 1950s by radar engineers. Unfortunately, received theoretical models of glint (or of scintillation for that matter) are static models, whereas the phenomenon is obviously dynamic when a radar system is tracking a moving target. These models are typically treated stochastically – e.g. time averages, serial correlations, cross spectra, etc. are calculated, when in fact the models do not evolve randomly over time. Randomness enters received theory in an ad hoc manner through arbitrary probability distributions that are assumed to describe the amplitudes, phases, and locations of the specular points which make up the target.[2] However, with a realization of these random variables in hand, the process becomes deterministic, or singular, even though it may appear to evolve unpredictably over time.

As a simple example of this point suppose a_i represents a set of random variables, and $b_i(t)$ a set of known functions of time. Then define the process,

$$x_t = \sum_{i=1}^{n} a_i b_i(t). \tag{7.1}$$

Equation (7.1) is an example of *purely deterministic process*.

In what follows we will develop a glint noise model which is also a proper stochastic process. The bicovariance function of the model has many nonzero terms. This means that the process possesses serial dependence beyond that which can be detected through the spectrum or autocorrelation function.

Linear Target

Ostrovityanov and Basolov (1985), pages 33-38, develop an expression for angle glint in the context of a tracking radar. Their model will serve as the starting point for developing a stochastic model of angle glint. The geometric set up for the problem is shown in Figure 7-1. Specular points are continuously distributed along the target's linear length ℓ. The points are statistically independent of one another. Assume that the length ℓ is small compared to its range, r, from an observer located at O. The apparent radar center is located at AC, and produces a glint error equal to angle φ. The orientation, or aspect, angle of the target is the angle q. The location of the specular points is defined in terms of their distances, ℓ_i, from the geometric center of the target, O1.

The total backscattered radar signal at the receiver, e_t^s, is at frequency f modeled as the sum of the reflections from the individual specular points:

$$e_t^s = \sum_i E_i \cos[\omega t - (\omega/c)\ell_i \sin q - \delta_i] \tag{7.2}$$

where the sum is taken over m specular points and $\omega = 2\pi f$. In (7.2) the E_i are amplitudes and the δ_i phases of the specular points, and c is the speed of light. Now let $\psi_i = (\omega/c)\ell_i \sin q$, and $\xi_i = \ell_i/(\ell/2)$, and define the four functions u_n, v_n, u_g, v_g, as follows:

Figure 7-1 Geometric of Linear Target at Range r from an Observer at Point O

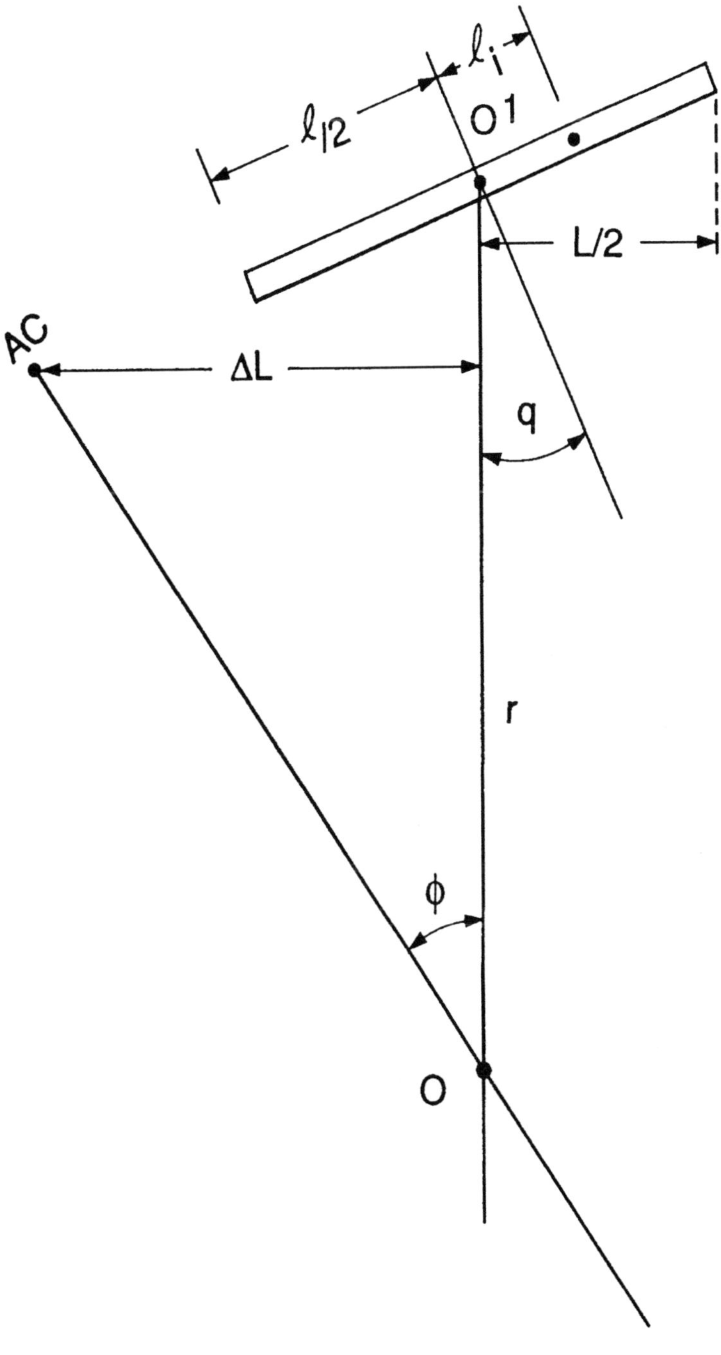

$$u_n = \sum_i E_i \cos\psi_i$$

$$v_n = \sum_i E_i \sin\psi_i$$

$$u_g = \sum_i \xi_i E_i \cos\psi_i$$

$$v_g = \sum_i \xi_i E_i \sin\psi_i \qquad (7.3)$$

Ostrovityanov and Basalov show that the linear error distance, ΔL, can be expressed as:

$$\Delta L = \frac{u_n u_g + v_n v_g}{u_n^2 + v_n^2}\left[\frac{L}{2}\right]. \qquad (7.4)$$

A similar expression obtains for the normalized range scintillation error. It is quite possible, as illustrated in Figure 7.1, for the apparent target center, AC, to be located off from the physical target.

The introduction of stochastic behavior to the model is greatly simplified, without losing the essence of glint errors, by assuming that the target consists of only two specular points. That is, while the m-point linear model provides better fidelity for relatively complex structures, the two-point model will more clearly highlight the underlying physics of glint as it randomly evolves through target motion. The new set up is illustrated in Figure 7-2.

Two-point Target

Equation (7.4) can be used to study the angle error (glint) of the two-point target by setting $\xi_{1,2} = \mp 1$ and rewriting equations (7.3):

$$u_n = E_1\cos\psi_1 + E_2\cos\psi_2$$

$$v_n = E_1\sin\psi_1 + E_2\sin\psi_2$$

$$u_g = -E_1\cos\psi_1 + E_2\cos\psi_2$$

$$v_g = -E_1\sin\psi_1 + E_2\sin\psi_2 \qquad (7.5)$$

Substituting the above equations into equation (7.4) yields the linear glint error for the two-point target:

$$\Delta L = \frac{E_2^2 - E_1^2}{E_2^2 + 2E_1E_2\cos\psi + E_1^2}\left[\frac{L}{2}\right]. \tag{7.6}$$

with $\psi = \psi_2 - \psi_1 = (\omega/c)2\ell \sin q + \delta$, being the phase difference between the two points. Note from Figure 7.2 that the target is oriented at angle q with respect to an observer at O. Equation (7.6) can be simplified by defining A as the ratio of the two amplitudes, $A = E_1/E_2$. Then

$$\Delta L = \frac{1 - A^2}{1 + 2A\cos\psi + A^2}\left[\frac{L}{2}\right]. \tag{7.7}$$

If r is the range to the target, the glint angle error, φ, is written

$$\varphi = \tan^{-1}(\Delta L/r). \tag{7.8}$$

The equation shows that the linear glint error depends on the ratio of the two amplitudes of specular points 1 and 2, as well as the phase angle between them. With $A = 0$ (i.e. $E_1 = 0$) the target consists of only specular point 2, $\Delta L = L/2$, and the tracking is at point 2. With $A = 0$, E_2 is 0 and the target consists of only specular point 1, $\Delta L = -L/2$, and that radar tracking is at point 1. With $A = 1$, and $\psi = 0$, then $\Delta L = 0$ and the tracking point is at the geometric center of the target. More generally, the ratio $\Delta L/(L/2)$ may assume any value between $-\infty$ and ∞, and the apparent target center may be located far off its physical boundary in either direction.

In order to simplify the notation, we introduce the normalized glint error, η, which is defined as

$$\eta = \frac{\Delta L}{L/2}. \tag{7.9}$$

Figure 7-2 Geometry of the Two-Point Target.

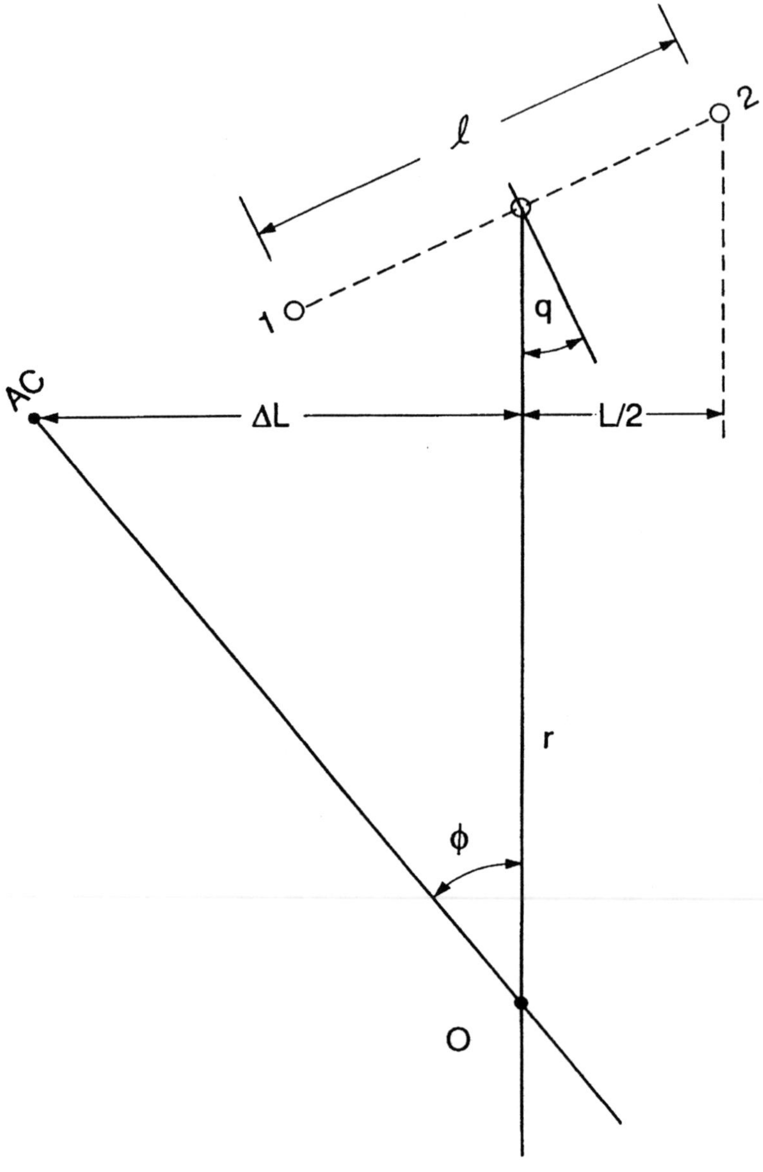

Quadratic Glint Model

Angle glint will tend to be greatest when $\psi \neq 0$ and A is close to one. In order to better understand the behavior of ΔL, or η, in dynamic situations, expand equation 7.7 as a Taylor series around the point ($A = h$, $\psi = k$). In particular, when the expansion is up to and includes terms of order 2, we have

$$\eta = [a_1(A - h) + a_2(\psi - k)] + \frac{1}{2}[b_1(A - h)^2 + b_2(A - h)(\psi - k) + b_3(\psi - k)^2] + R. \tag{7.10}$$

The remainder term is of order ⅓!. The coefficients in (7.10) are denoted by lower case subscripted letters and are found by evaluating the partial and cross-partial derivatives of equation (7.7) at $A = 0.95$ and $\psi = \pi/4$. For example, in the quadratic case of (7.10):

$$\eta = -0.374(A - 0.95) + 0.12(\psi - \pi/4) + \tfrac{1}{2}[0.731(A - 0.95)^2 - 0.193(A - 0.95)(\psi - \pi/4) + 0.010(\psi - \pi/4)^2] + R. \tag{7.11}$$

Next, consider the random time-series properties of glint error. The only sensible source of stochastic behavior is through the target orientation angle, q, which in turn makes ψ stochastic. The behavior of q, now written q_t, over time could manifest itself either as a series of independent random movements, or the movements could be correlated. Which case is most relevant depends upon the nature of the specific target: it might be stationary, but subject to random buffeting by the wind, or the target could be a turning aircraft subject to small random changes in velocity and heading. In the event, because q_t is the source of the stochastics in the model, we will treat the $(A - h)^n$ terms in Equation (7.10) as constants that are time invariant, and regard the phase difference ψ_t of order 2:

$$\eta_t = a_0 + a_1(\psi_t - k) + a_2(\psi_t - k)^2 + 0, \tag{7.12}$$

with

$$\psi_t = \left(\frac{\omega}{c}\right) 2\ell \sin q_t - \delta, \tag{7.13}$$

and note that the remainder term in (7.11) has been replaced by a random order term in (7.12) of order 1/(3!).

Assume that q_t is generated by an autoregressive time series model of order p. This AR(p) model is written as:

$$q_t = \varphi_1 q_{t-1} + \varphi_2 q_{t-2} + \dots + \varphi_p q_{t-p} + e_t. \qquad (7.14)$$

The innovation series, e_t, is assumed to be i.i.d. normal with zero mean and standard deviation σ_e. Furthermore, let σ_e be small, i.e. $\sigma_e << 1$ radian. Then q_t will be small and $\sin q_t \approx q_t$. Hence, ψ_t will approximately follow an AR(p) process with a nonzero mean:

$$\psi_t \simeq \mu + \varphi_1 \psi_{t-1} + \varphi_2 \psi_{t-2} + \dots + \varphi_p \psi_{t-p} + e_t. \qquad (7.15)$$

Below it will be assumed that the approximation in (7.15) holds exactly, which implies that ψ_t is distributed as a normal random variable. Given this relationship between ψ_t and its past values, it is apparent that the expression for η_t in (7.12) involves lagged cross products of ψ_t.

Next, assume that the approximation in (7.12) holds exactly, and rewrite the equation as

$$\eta_t = a_0 + a_1 k^2 - a_1 k + (a_1 - 2a_2 k)\psi_t + a_2 \psi_t^2. \qquad (7.16)$$

Substitute (7.15) into (7.16):

$$\begin{aligned}
\eta_t &= a_0 + a_1 k^2 - a_1 k \\
&+ (a_1 - 2a_2 k)(\mu + \varphi_1 \psi_{t-1} + \varphi_2 \psi_{t-2} + \dots + \varphi_p \psi_{t-p} + e_t) \\
&+ a_2[\mu^2 + \mu\varphi_1 \psi_{t-1} + \dots + \mu\varphi_p \psi_{t-p} + \mu e_t \\
&+ \mu\varphi_1 \psi_{t-1} + \varphi_1^2 \psi_{t-1}^2 + \dots + \varphi_1 \varphi_p \psi_{t-1} \varphi_{t-p} + \varphi_1 \psi_{t-1} e_t \\
&+ \mu\varphi_2 \psi_{t-2} + \varphi_1 \varphi_2 \psi_{t-1} \psi_{t-2} + \dots + \varphi_2 \varphi_p \psi_{t-2} \psi_{t-p} + \varphi_2 \psi_{t-2} e_t \\
&\vdots \\
&+ \mu e_t + \varphi_1 \psi_{t-1} e_t + \dots + \varphi_p \psi_{t-p} e_t + e_t^2].
\end{aligned} \qquad (7.17)$$

The above expression for η_t is an example of a nonlinear time series of the quadratic type with a correlated linear term. There are two sources of serial dependence in equation (7.17): the linear lag terms of form $\varphi_i \psi_{t-i}$ which imply autocorrelation in η_t, and the second-order terms such as the lagged cross product $\varphi_1 \varphi_2 \psi_{t-1} \psi_{t-2}$ which imply a nonzero bicovariance function for η_t. The product terms found in (7.17) also indicate that η_t is not normally distributed, although ψ_t is assumed to be a normal random variable.

The bicovariance function of η_t, also known as the third-order cumulant function, is defined as

$$C_{\eta\eta\eta}(\theta,\tau) = E[\eta_t - \mu][\eta_{t+\theta} - \mu][\eta_{t+\tau} - \mu]. \tag{7.18}$$

Evaluating (7.18) using the expression for η_t in (7.17) would prove to be very cumbersome; at this point it is more important to realize that $c_{\eta\eta\eta}(\theta, \tau)$ is nonzero for many values of θ, $\tau > 0$.

Cubic Glint Model

A cubic model for glint can be developed by including terms of order 3 in the power series expansion of equation 7.10:

$$\begin{aligned}\eta &= [a_1(A - h) + a_2(\psi - k)] \\ &+ \frac{1}{2}[b_1(A - h)^2 + b_2(A - h)(\psi - k) + b_3(\psi - k)^2] \\ &+ \frac{1}{3}[c_1(A - h)^3 + c_2(A - h)^2(\psi - k) \\ &+ c_3(A - h)(\psi - k)^2 + c_4(\psi - k)^3] + R.\end{aligned} \tag{7.19}$$

where the remainder term is now of order 1/(4!). Again, the phase difference, ψ_t, is the source of the stochastics, and η_t can therefore be expressed as a cubic equation in ψ_t. Hence, after dropping the remainder term we have the time series model:

$$\eta_t = a_0 + a_1(\psi_t - k) + a_2(\psi_t - k)^2 + a_3(\psi_t - k)^3. \tag{7.20}$$

Clearly, if (7.15) were substituted into equation (7.20), we would have a time series model with linear terms and second-order terms as in (7.17) above, plus lagged cubic product terms such as $\varphi_i\ \varphi_j\ \varphi_k\ \psi_{t-1}\psi_{t-j}\ \psi_{t-k}$. Thus, we can model the glint noise as a nonlinear time series of the cubic type with a correlated linear term.

The tricovariance function is analogous to the bicovariance function considered under the quadratic model. It is defined as

$$\begin{aligned}c_{\eta\eta\eta\eta}(\theta,\tau,\alpha) &= M_{\eta\eta\eta}(\theta,\tau,\alpha) - c_{\eta\eta}(\tau)c_{\eta\eta}(\theta - \alpha) \\ &- c_{\eta\eta}(\alpha)c_{\eta\eta}(\theta - \tau) - c_{\eta\eta}(\theta)c_{\eta\eta}(\tau - \alpha)\end{aligned} \tag{7.21}$$

where $M_{\eta\eta\eta}(\theta,\tau,\alpha) = E[\eta_t - \mu][\eta_{t+\theta} - \mu][\eta_{t+\tau} - \mu][\eta_{t+\alpha} - \mu]$ and $c_{\eta\eta}(\bullet)$ is the covariance function for η_t. It can be shown using (7.20) above that many of the fourth-order cumulant terms will also be nonzero for values of θ, τ, $\alpha > 0$. This is equivalent to saying the η_t has a nonzero trispectrum.

Finally, note that deconvolving η_t in order to remove the autocorrelation will *not* remove the third- and higher-order dependence in the glint signal. Hence, the use of linear filtering methods, such as deconvolution, will not provide an optimum solution to detection, tracking, classification, and localization problems. In general, signal processing and beam forming methods which take into account the higher order dependence in glint and scintillation errors will provide superior results.

ANALYSIS OF GLINT NOISE

The data comes from a radar cross-section measurement range operated by the U.S. Navy. The target is a model of a T-38 jet trainer. The measurements were calibrated such that the target was in the resonance region of the radar, that is, the wavelength is on the order of the target size. The target was mounted on a rotating pedestal; as the pedestal turned, the azimuth angle to the apparent target center was measured every 0.1 degree of rotation. The total number of observations is 3600 (0.1° x 360°). In this analysis we use the first 2000 observations, which correspond to a rotation of 200 degrees relative to a reference point where the nose of the aircraft is pointed directly at the radar. A plot of the data is displayed in Figure 7-3. A horizontal reference line is drawn at 180 degrees, the correct azimuth angle to the target. The mean error in angle is -1.075 degrees with a standard deviation of 14.6 degrees. The data has little skewness (-0.75), but is highly leptokurtic; the kurtosis is 11.83.

By the nature of the experimental design, the glint error sequence is a purely deterministic process. As a consequence, we feel it is not appropriate to treat the data as a realization of a stochastic process.[3] Instead, we use the glint angle errors as the basis of a simulation of a glint noise stochastic process. The aspect angle, q, is generated from a stationary AR(2) model ($\varnothing_1 = 0.50$, $\varnothing_2 = 0.45$) with the standard deviation of the innovation set to 5 degrees; this model is driven by pseudo-random Gaussian innovations. The glint error for each q is obtained by looking it up in the original data set shown in Figure 7-3. Thus, we generate a simulated stochastic process of 2000 observations of the glint error from a T-38 target when it is being stochastically perturbed from a mean rotation of 100 degrees.

A realization of the simulation is plotted in Figure 7-4. As the figure shows, the angle error is offset from zero by a positive amount; the sample mean of this series is 5.7 degrees. The other sample statistics are: standard deviation, 15.6 degrees, skewness 0.64, and kurtosis 4.06. The maximum and minimum glint errors are +64 and -62 degrees, respectively.

It is well known that the power spectrum of glint noise data contains a dominant low-frequency component. This is quite visible in Figure 7-5 where we display the logarithm of the spectrum of the simulated glint noise data. The low frequency component is called *bright-spot wander*, and represents the electromagnetic center of the target. The low energy peak in the power spectrum means that the

Figure 7-3 Plot of Glint Angle Errors Measured Using A Model of a T-38 Jet Trainer

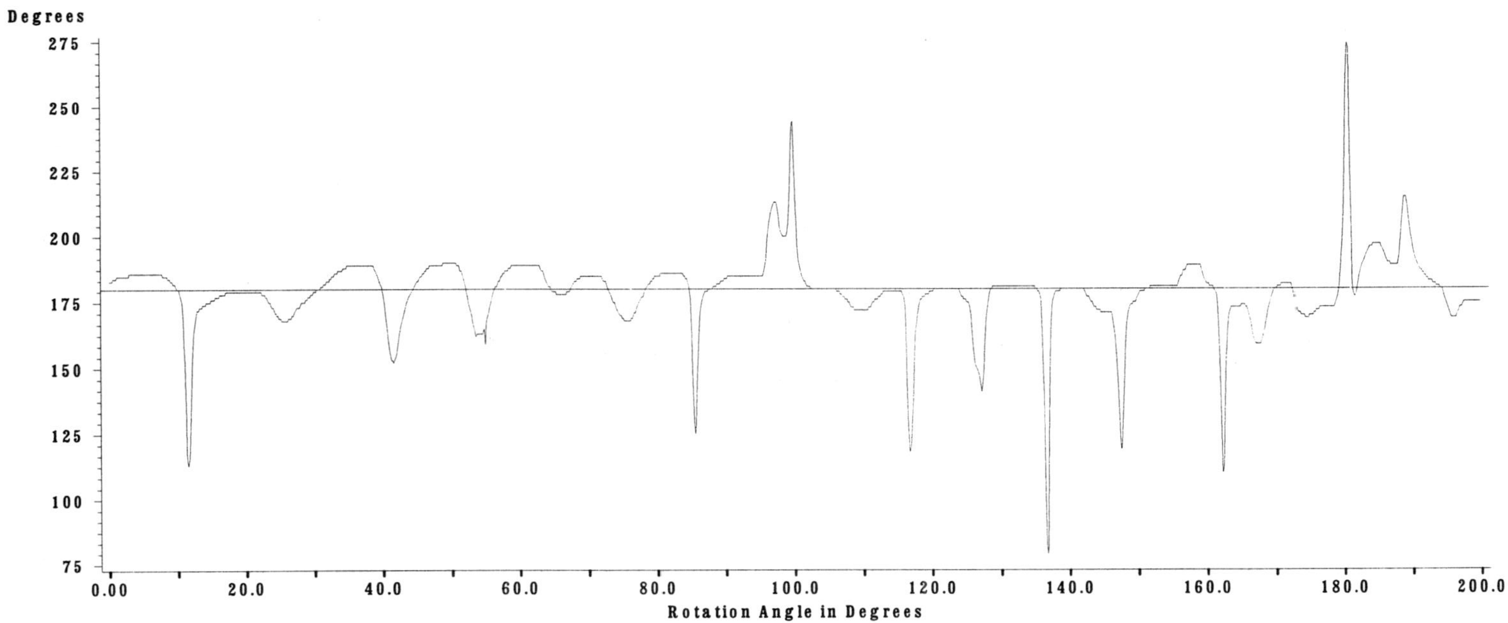

Figure 7-4 Plot of Simulated Glint Noise Generated from Data Shown in Figure 7-3

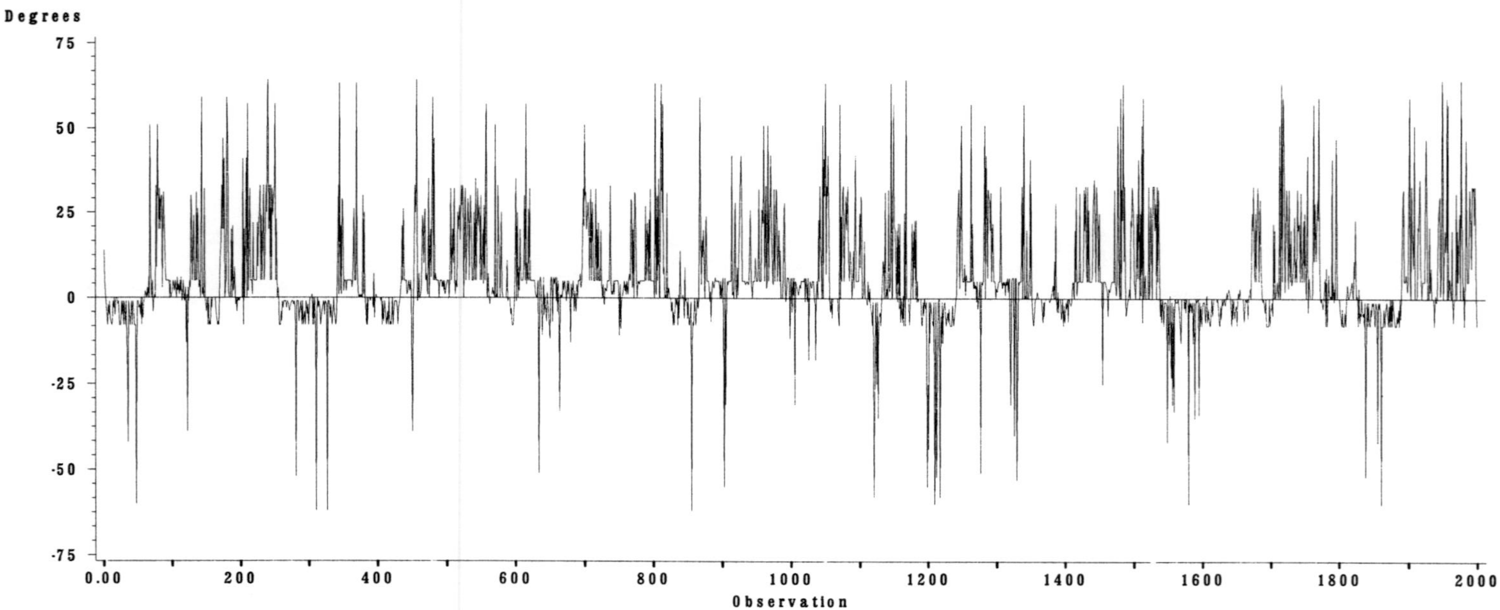

Figure 7-5 Power Spectrum of the Simulated Glint Noise Process

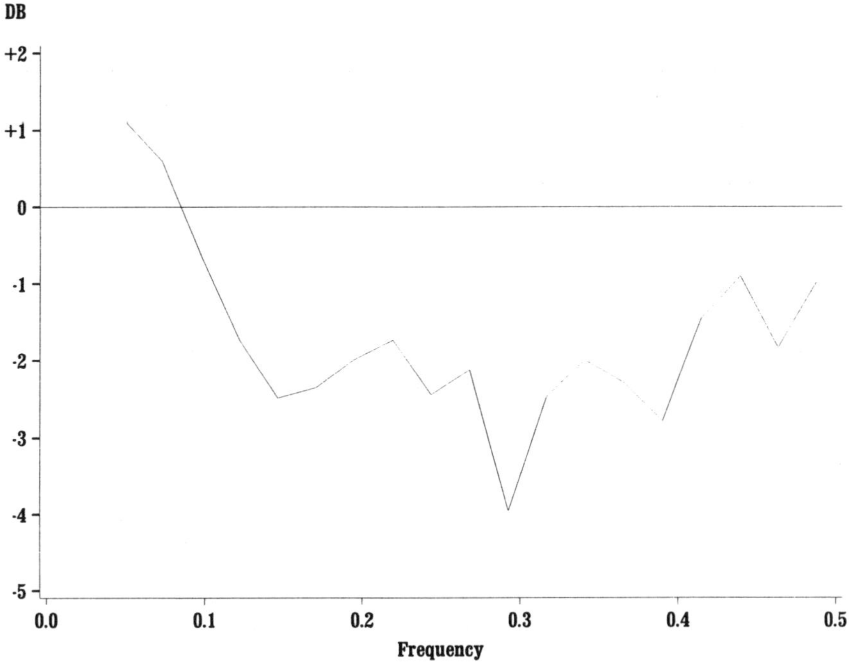

process is not white noise, and will need to be prewhitened before it is presented to our battery of nonlinearity tests. According to the Schwartz Criterion, an AR(5) model is best for prewhitening the simulation data in Figure 7.4.

The results from running the nonlinearity tests on the glint noise simulation data are displayed in Table 7-1. We see from the Table that not a single test based upon third-order moments (Tsay, Bicovariance, or Bispectral) is significant at either the 5% or 10% level. The result is not particularly surprising given the low skewness and kurtosis in the data. Nevertheless, there is strong evidence of nonlinearity provided by the remaining tests. Apparently the fourth-order cumulant terms suggested by Equation (7.20) are more important than the quadratic terms from Equation (7.17). Furthermore, Figure 7-4 indicates apparent bursts of volatility in the data. This suggests that an ARCH-type model might be quite useful in modeling the nonlinear dependence. The evidence presented here strongly suggests that in order to adequately filter or predict glint noise errors one must take into account the nonlinear serial dependence in the underlying physical process.

Table 7-1 P-Values for Nonlinearity Tests on Glint Noise Simulation[1,2]

Test	P-Value
McLeod-Li	0.047
Engle LM	0.024
BDS	
m = 2	0.00
m = 3	0.00
m = 4	0.00
Tsay	0.357
Bicovariance	0.254
Bispectral	0.226

1. Parameter values for the tests were set as follows: McLeod-Li 20 lags, Engle 8 lags, BDS ε=1.0, Tsay 10 lags, Bicovariance 20 lags, Bispectrum smoothing window 48.

2. Test statistics computed using 1000 bootstraps and 2000 observations.

Notes

1. This chapter draws heavily from D.M. Patterson (1991) "Third and Higher-Order Dependence in Active Sonar Tracking Errors," FINAL REPORT, Prepared for Naval Underwater Systems Center, Under Contract No. N70024-0193-0441.

2. See for example Delano (1953) and Muchmore (1960).

3. Martin and Thomson (1982) analyze glint data that was produced in a similar manner to our data. Their interest was spectral estimation. However, they either did not realize, or were unconcerned, that the data is deterministic.

REFERENCES

Delano, R. (1953) "A Theory of Target Glint or Angular Scintillation in Radar Tracking," *Proceedings of the I.R.E.* 41, 1778-1784.

Martin, R. and D. Thompson (1982) "Robust – Resistent Spectrum Estimation," *Proceedings of the IEEE* 70, 1097-1115.

Muchmore, R. (1960) "Aircraft Scintillation Spectra,"*IRE Transactions on Antennas and Propagation* AR-8(2), 201-212.

Ostrovityanov, R. and F. Basalov (1985) *Statistical Theory of Extended Radar Targets*, translated by W. Barton and D. Barton, Artech House, Inc., Dedham, MA.

8 SEISMIC DATA

INTRODUCTION

Nowadays earthquakes are well monitored, yielding copious amounts of seismic time series data. In this Chapter we examine seismic data from a recent California earthquake for evidence of nonlinearity in its generating mechanism.

DATA

Digitized strong-motion accelerograms were obtained from the United State Geological Survey National Strong-Motion Program (http://nsmp.wr.usgs.gov/) for the 1992 Landers, California earthquake. This is the first strong earthquake to occur in the contiguous United States for which good data exists.

The Landers earthquake took place on June 28, 1992. Registering 7.6 on the Richter scale, it is estimated to be the ninth strongest earthquake to hit the contiguous United States in modern times. It caused three deaths, more than 400 injuries, and over $92 million in damage. Locally, surface fault displacements in excess of 5.5 meters horizontally and 1.8 meters vertically were observed; the shock was felt throughout Southern California and portions of the surrounding states.

Uncorrected acceleration data is available for the Landers quake from seven recording stations. The data analyzed here was measured 43 kilometers away from the epicenter at the Thousand Palms post office (station #5068; recorder #5304) on channel1, which registered acceleration in a horizontal orientation 135° east of north – i.e., in the SE-NW direction. “Uncorrected” in this context means that the data has not been filtered to eliminate “noise” or frequency-dependent instrument responses; it has, however, been corrected for uneven film transport through the recorder and translated from recorder units into cm/sec^2.

This data set contains 11,383 acceleration observations, digitized from the original recorder film at a rate of 200 samples per second. Thus, the entire record corresponds to a time period of just under 57 seconds.

The complete data set is plotted in Figure 8-1. Since this series appears to be nonstationary, we next considered a subset of the data, from observation number 1000 to observation number 7000; these data are plotted in Figure 8-2 and appear to be reasonably stationary over this restricted period.[1]

Following Box and Jenkins (1976) and Granger and Newbold (1977), we next examined the sample autocorrelation function of the data over this sample period; these results are given in Table 8-1 and plotted in Figure 8-3 below.[2] Since the sample autocorrelation of the accelerations at lag 1 is essentially one and decays quite slowly with increasing lag, we conclude that this series has a unit root and next examine the sample autocorrelation function of the differenced series, the change in the acceleration; these results are given in Table 8-1 and plotted in Figure 8-4 below. Since the sample autocorrelation of the change in the accelerations at lag 1 is substantially less than one and decays quickly with increasing lag, we conclude that further differencing is not necessary. A time plot of the change in the accelerations is given in Figure 8-5; it appears to be reasonably stationary over the period running from observation number 1000 to observation number 7000. Consequently, the nonlinearity tests are applied to the differenced acceleration data below.

NONLINEARITY TEST RESULTS

Table 8-2 summarizes the nonlinearity test results for the differenced acceleration data. Except for the bispectral test, all of the tests strongly reject the null hypothesis of a linear generating mechanism for this time series over the sample period running from observation number 1000 to observation number 7000.

These rejections are so strong, however, that they convey little information about which of the tests are most sensitive to the kind of nonlinearity in these data, so we also present results in Table 8-2 for a number of subperiods. The data for each subperiod considered is plotted in Figures 8-6 to 8-17 below. The period 1000 to 3000 was examined to eliminate the possibility that some of the larger observations observed in the period 1000 to 3000 and 6000 to 7000 might be responsible for the strong rejection of the null hypothesis of linearity over the period 1000 to 7000; evidently, they are not.

The remainder of Table 8-2 partitions the entire data set into segments of 1000 observations each; the shading in the Table is used to highlight the results for the subperiods which together constitute the 1000 to 7000 sample period. Several interesting conclusions emerge from this shaded portion of the Table. First, note that the evidence for a nonlinear generating mechanism for this time series is clearly much stronger in the latter portion of the sample, from observation 5000 to observation 7000. This suggests that the underlying character of the shaking was changing over this period. Yet an examination of Figure 8-6 shows that a time plot of the data from observation 3000 to observation 6000 gives no hint of this shift; evidently, the tests can detect variations in the intensity and/or kind of nonlinearity which are not apparent to the naked eye.

Figure 8-1 Landers, CA Earthquake Observed Accelerations
(All Data Units: cm/sec^2)

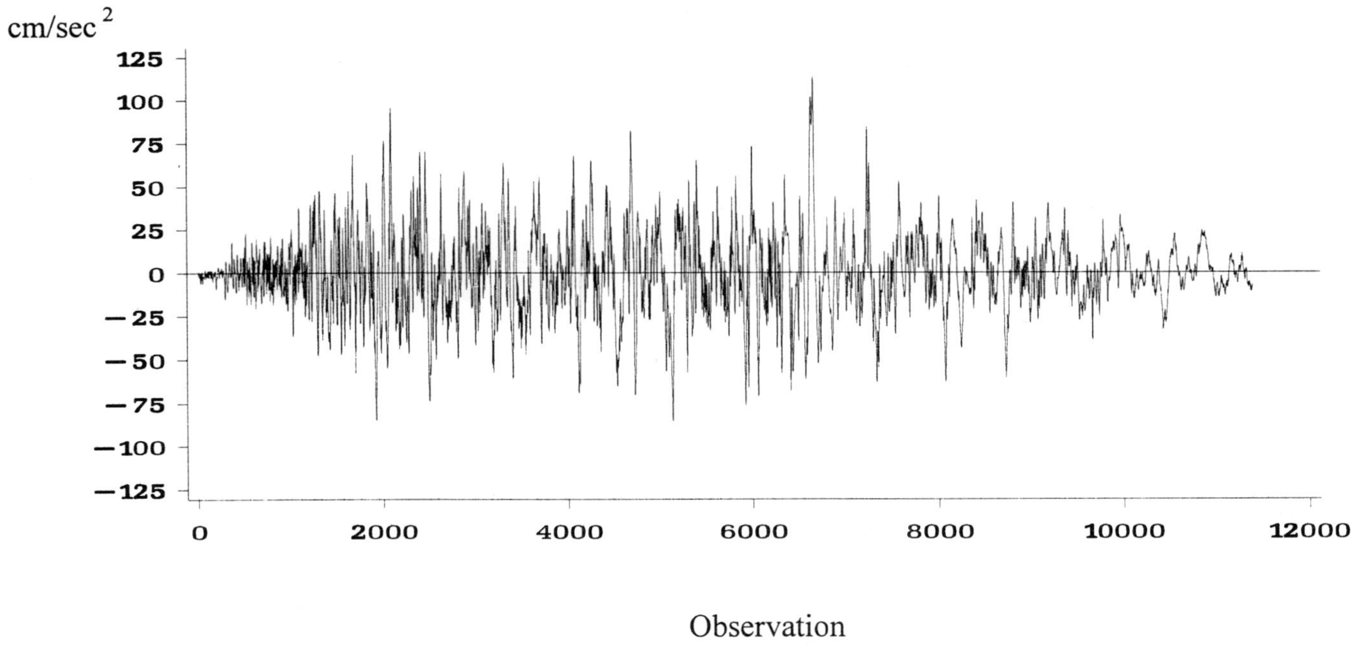

Figure 8-2 Landers, CA Earthquake Observed Accelerations
(Observations 1000-7000 Units: cm/sec ^{2})

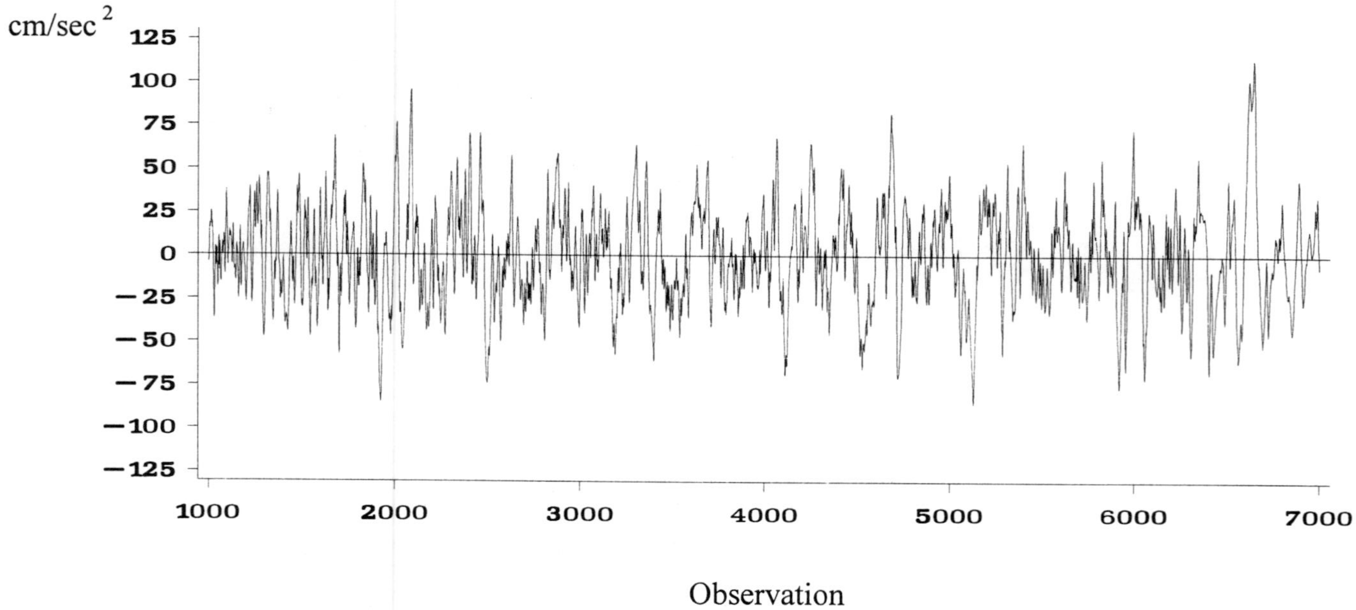

Figure 8-3 Landers, CA Earthquake Change In Acceleration
(Observations 1000-7000 Units: cm/sec^2)

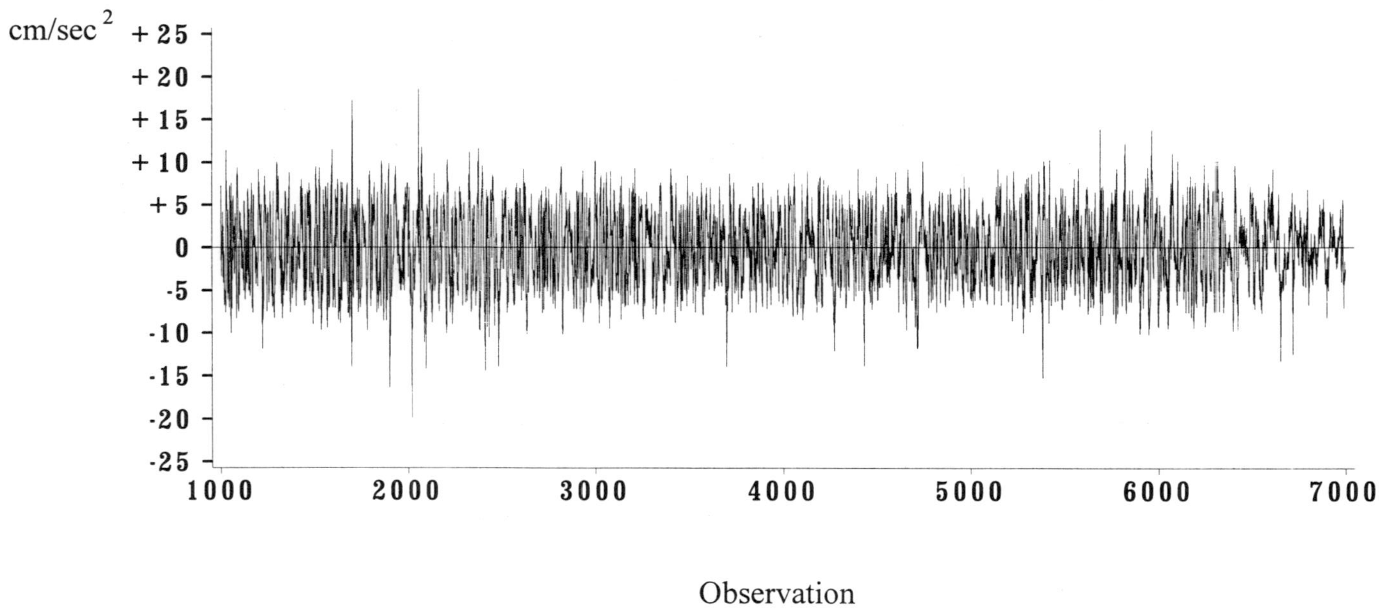

Table 8-1 Sample Autocorrelations
Landers Earthquake data[a]

lag	acceleration	change in acceleration
1	.988	.746
2	.959	.546
3	.917	.343
4	.866	.177
5	.811	.047
6	.756	-.040
7	.701	-.093
8	.649	-.113
9	.599	-.114
10	.552	-.111
11	.507	-.105
12	.465	-.010
13	.426	-.104
14	.389	-.116
15	.354	-.119

[a] This table lists the sample autocorrelations at lags 1, 2, 3, ... and so forth. Since the data is sampled from the seismograph film at 200 samples per second, each lag corresponds to .005 second. The sampling errors in these estimates are positively correlated due to the positive Autocorrelations at low lags.

Figure 8-4 Landers, CA Earthquake Sample Autocorrelations
(Acceleration All Data)

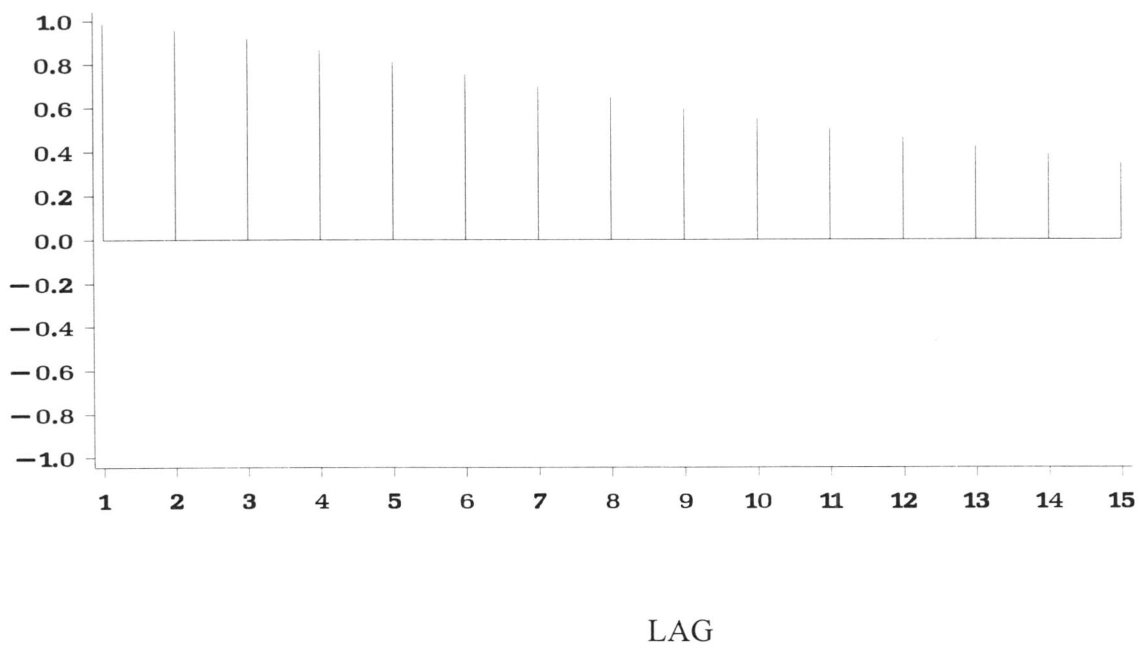

Figure 8-5 Landers, CA Earthquake Sample Autocorrelations
(Change In Acceleration All Data)

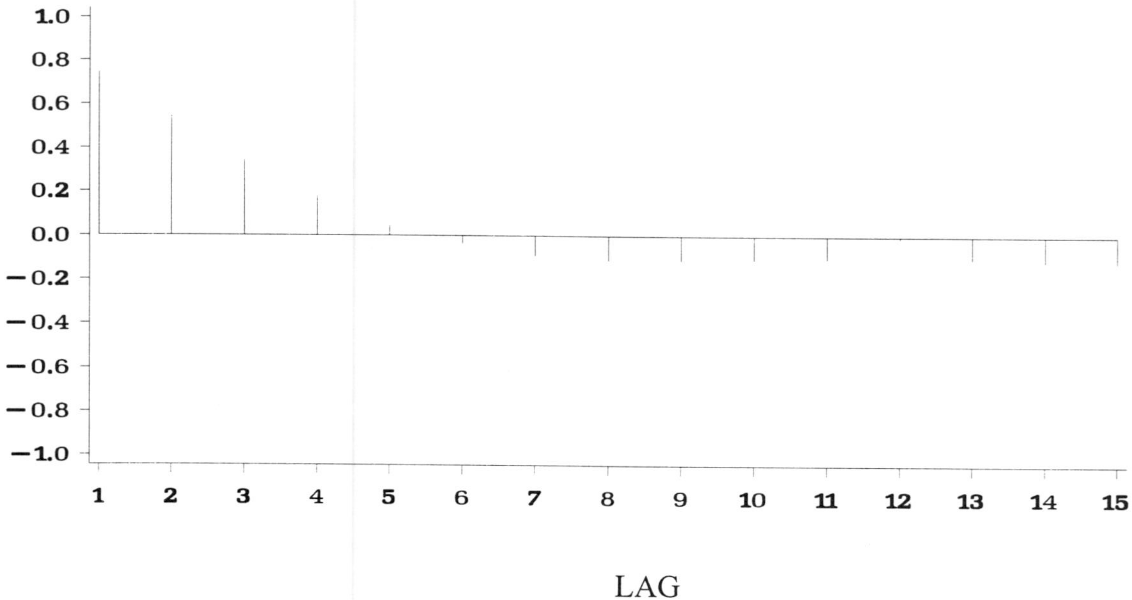

Table 8-2 Landers Earthquake Data
Nonlinearity Test Significance Levels[a]

Sample	McLeod-Li	Engle LM	BDS			Tsay	Bicov.	Bispectral
1000 - 7000	.00	.00	.00	.00	.00	.02	.00	.06
3000 - 6000	.00	.00	.00	.00	.00	.01	.00	.14
1- 1000	.00	.00	.00	.00	.00	.51	.00	.37
1000 - 2000	.80	.72	.86	.68	.21	.61	.63	.72
2000 - 3000	.04	.00	.16	.14	.13	.27	.20	.82
3000 - 4000	.56	.11	.66	.33	.05	.96	.25	.36
4000 - 5000	.01	.00	.02	.08	.07	.38	.16	.18
5000 - 6000	.00	.00	.00	.01	.00	.00	.00	.01
6000 - 7000	.00	.00	.00	.00	.00	.00	.00	.02
7000 - 8000	.00	.00	.00	.00	.00	.01	.16	.05
8000 - 9000	.00	.01	.02	.01	.02	.27	.01	.77
9000 - 10000	.02	.60	.43	.19	.10	.53	.18	.16
10000 - 11000	.42	.14	.27	.41	.65	.35	.18	.26

[a]Based on 1000 bootstrap samples. Each test is described in Chapter 2. Shaded rows correspond to stationary periods.

Figure 8-6 Landers, CA Earthquake Change In Acceleration
(Observations 3000-6000   Units: cm/sec^2)

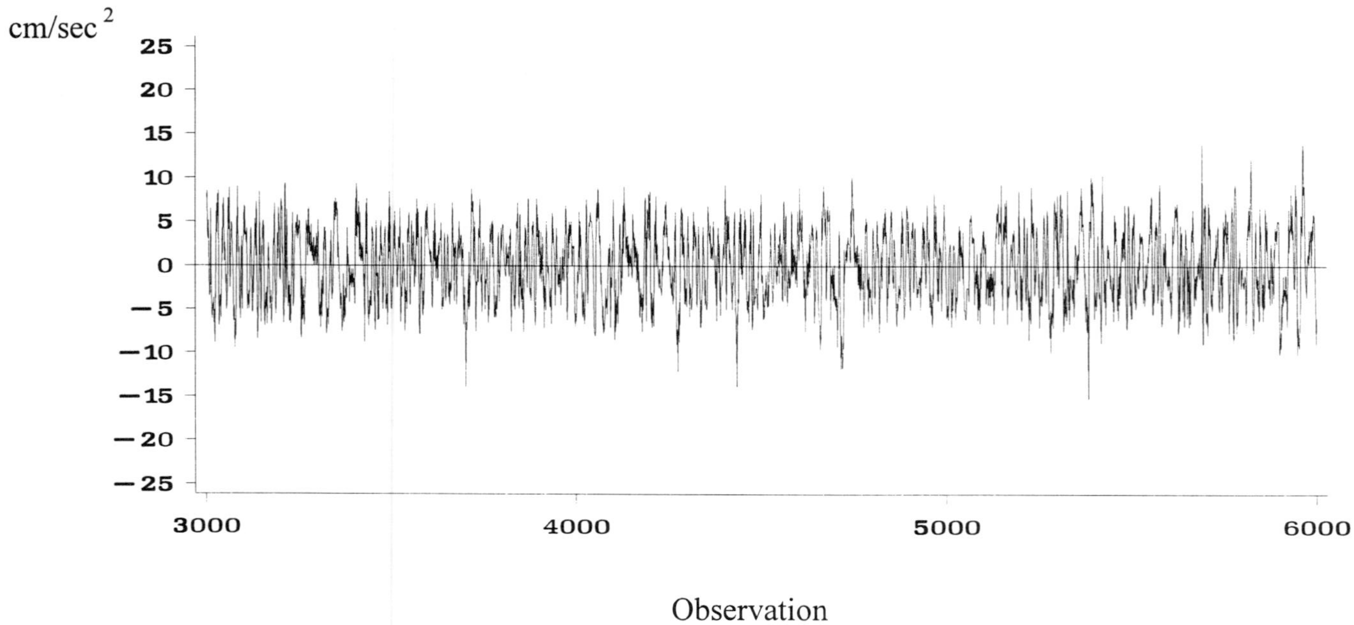

Figure 8-7 Landers, CA Earthquake Change In Acceleration
(Observations 0-1000   Units: cm/sec ^{2})

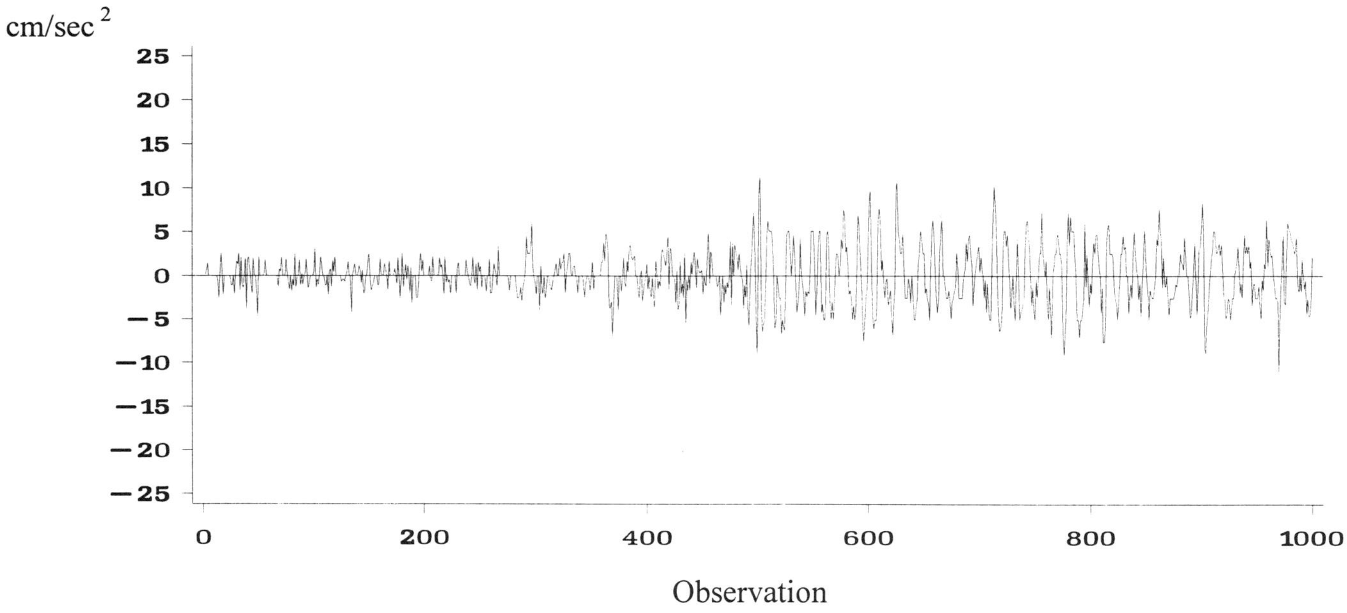

Figure 8-8 Landers, CA Earthquake Change In Acceleration
(Observations 1000-2000 Units: cm/sec^2)

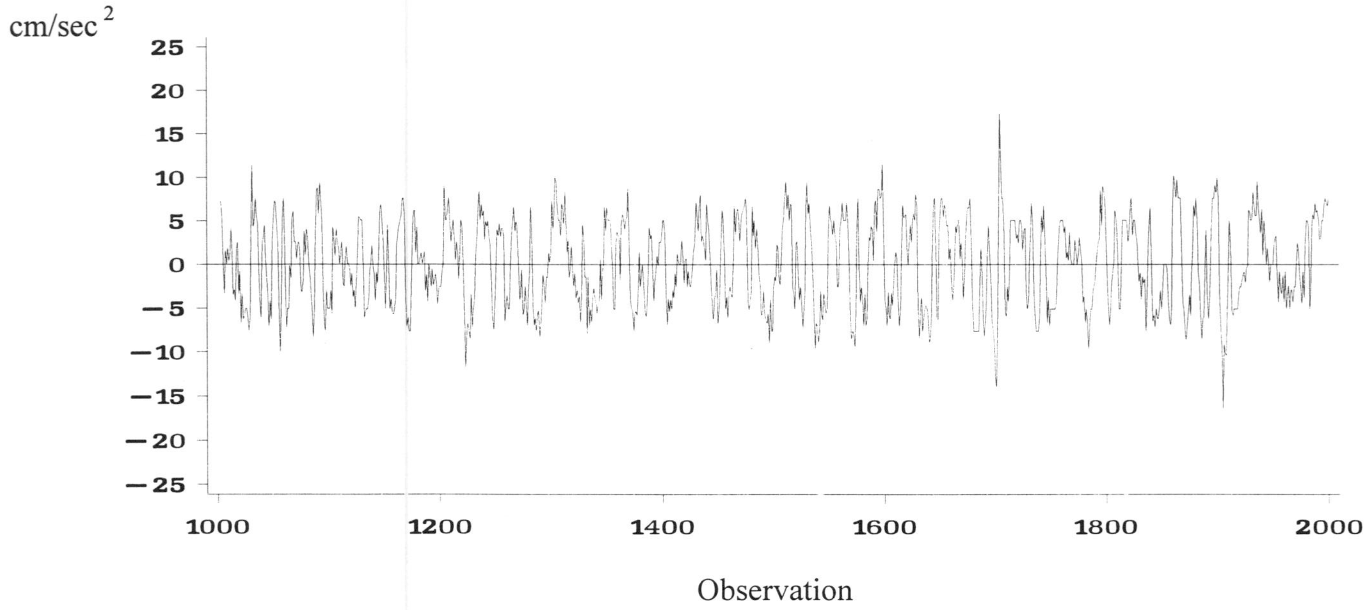

Figure 8-9 Landers, CA Earthquake Change In Acceleration
(Observations 2000-3000 Units: cm/sec^2)

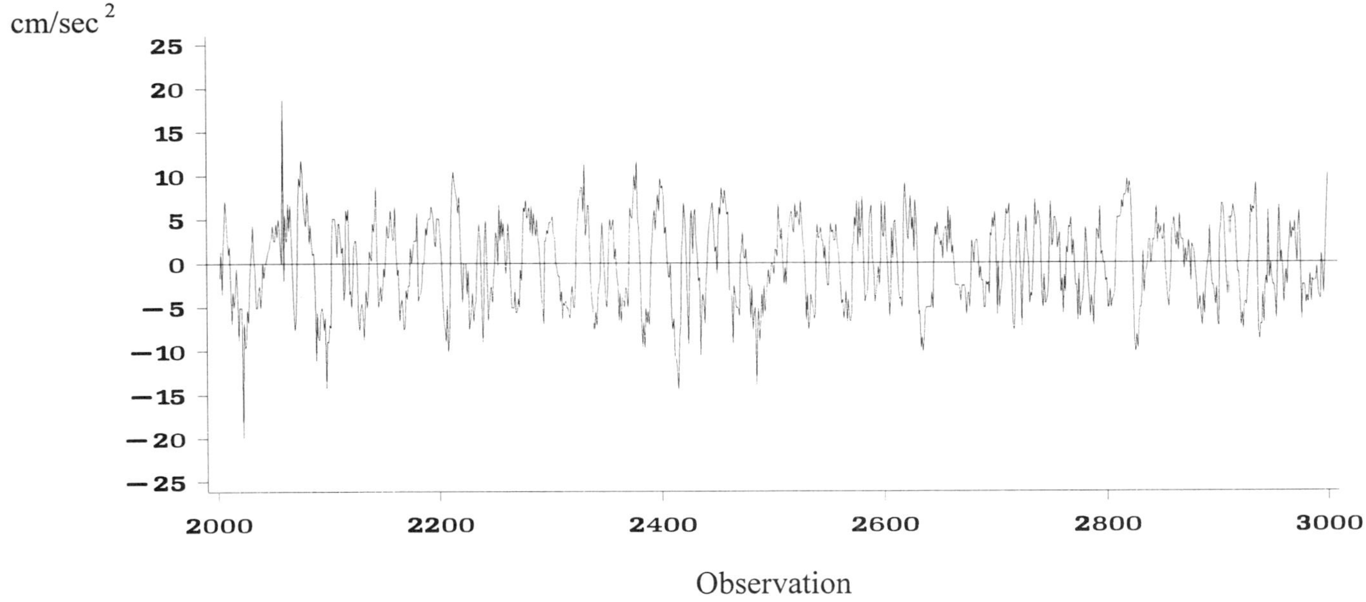

Figure 8-10 Landers, CA Earthquake Change In Acceleration
(Observations 3000-4000 Units: cm/sec ^{2})

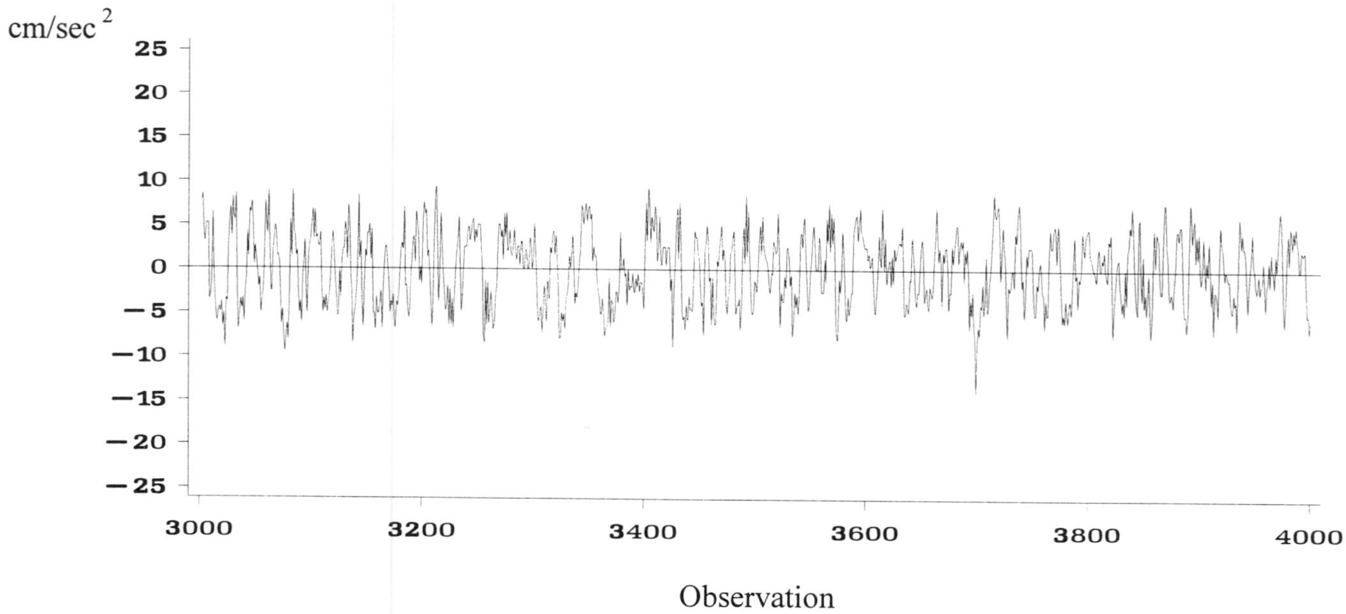

Figure 8-11 Landers, CA Earthquake Change In Acceleration
(Observations 4000-5000 Units: cm/sec^2)

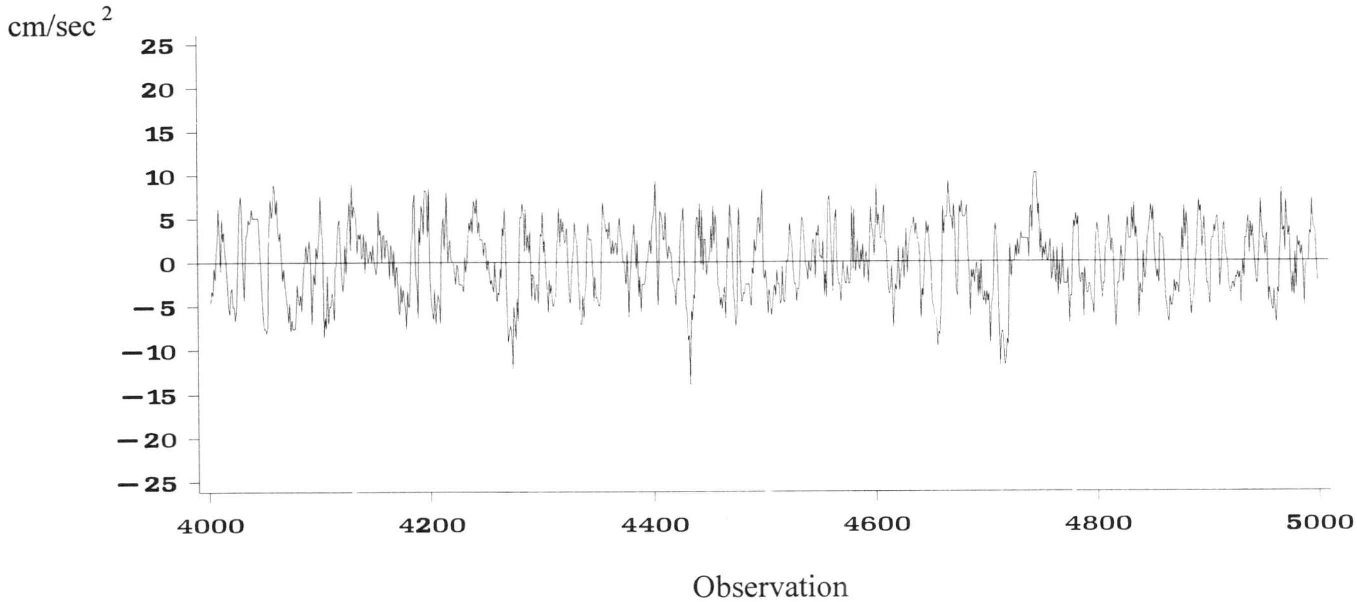

Figure 8-12 Landers, CA Earthquake Change In Acceleration
(Observations 5000-6000 Units: cm/sec ^{2})

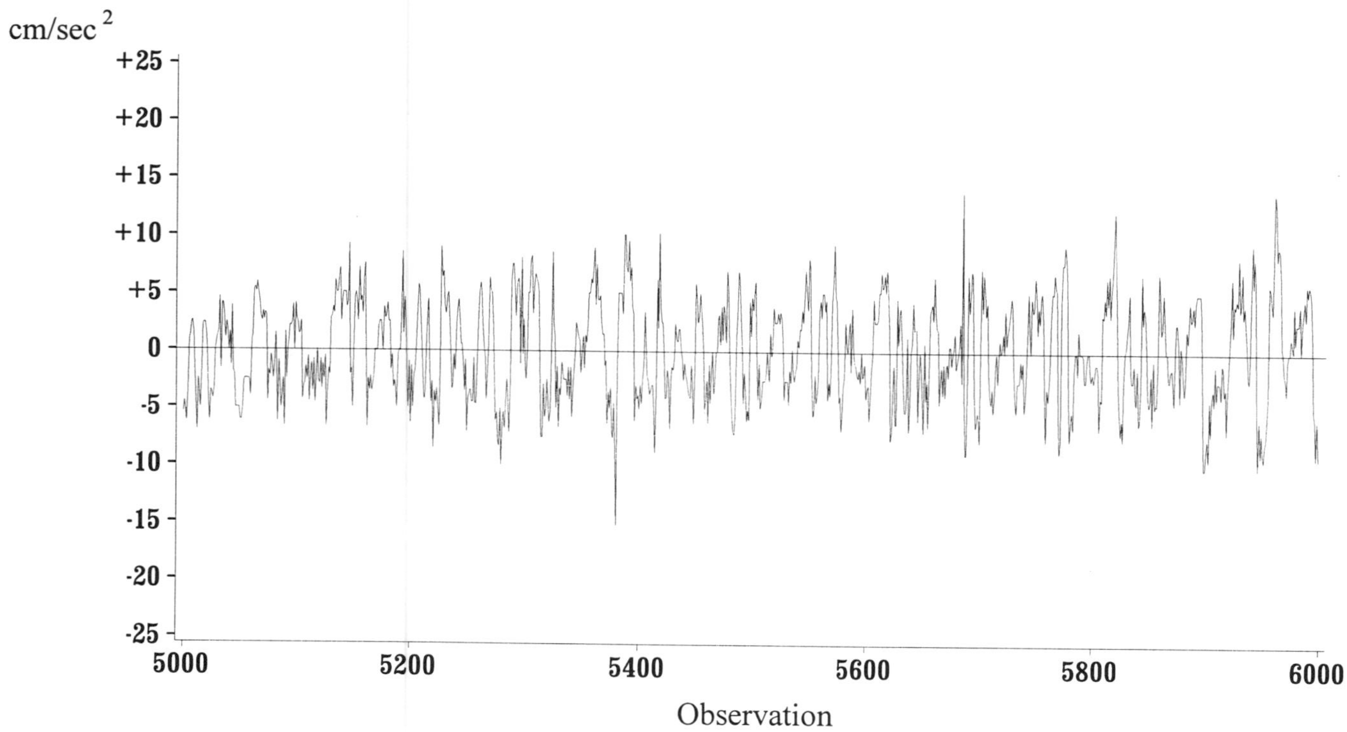

Figure 8-13 Landers, CA Earthquake Change In Acceleration
(Observations 6000-7000 Units: cm/sec ^{2})

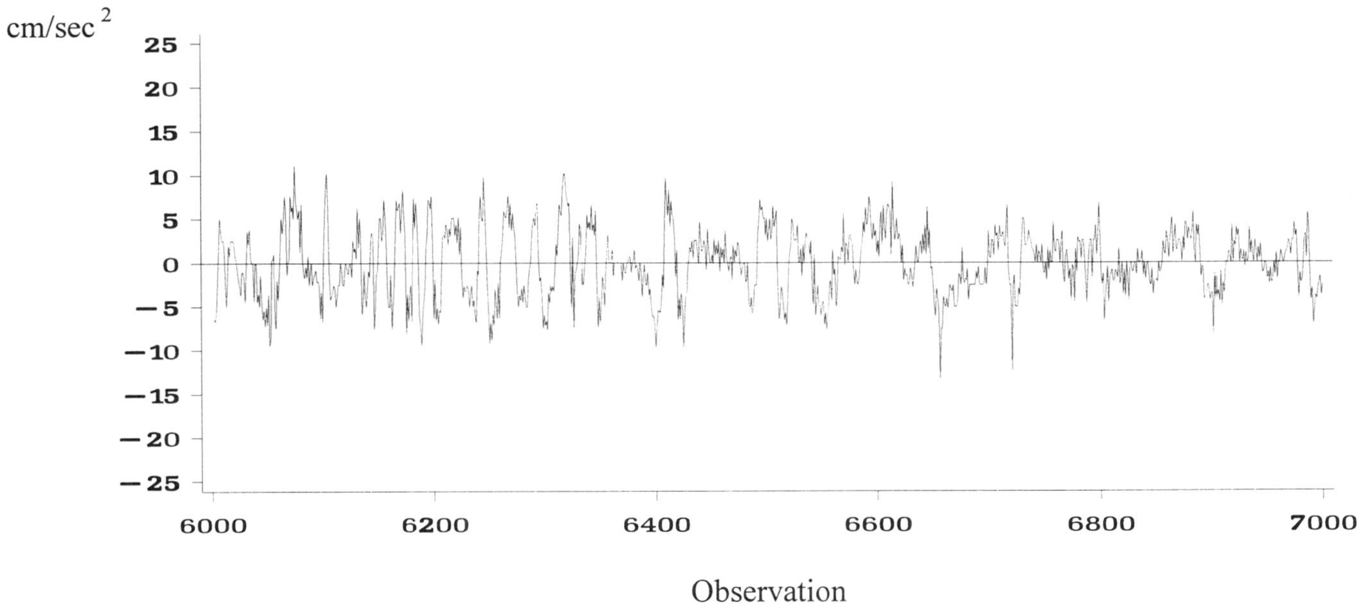

Figure 8-14 Landers, CA Earthquake Change In Acceleration
(Observations 7000-8000   Units: cm/sec^2)

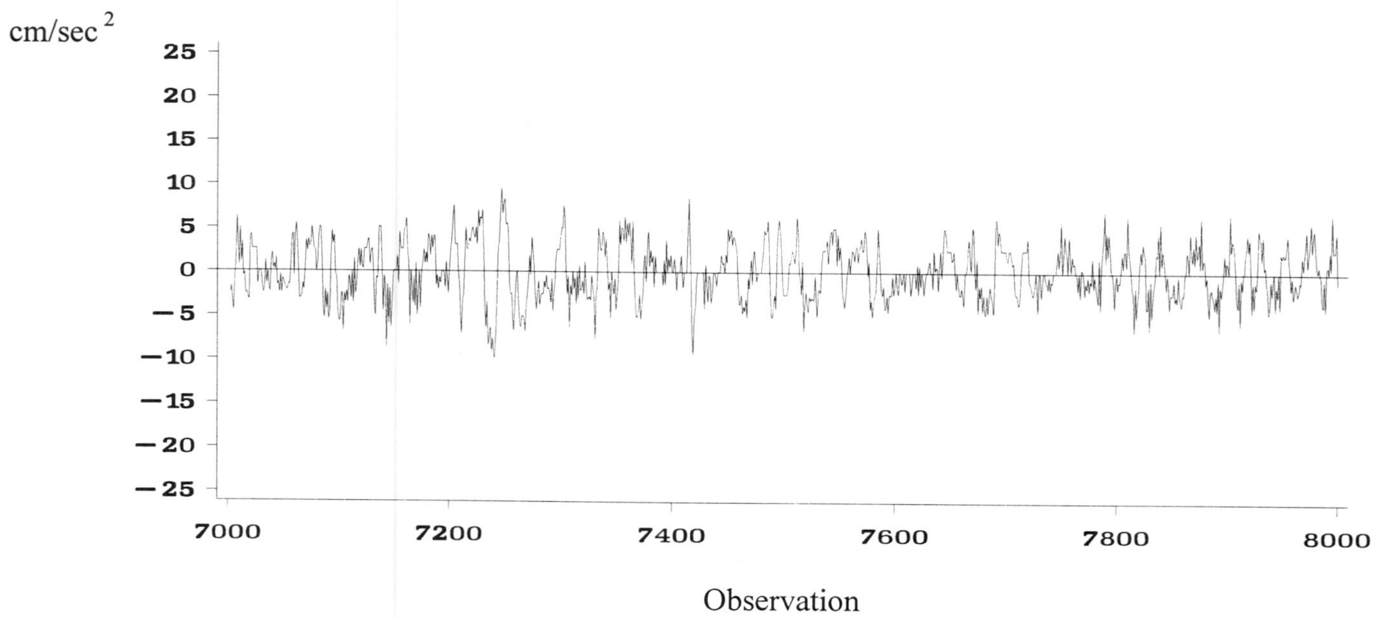

Figure 8-15 Landers, CA Earthquake Change In Acceleration
(Observations 8000-9000 Units: cm/sec^2)

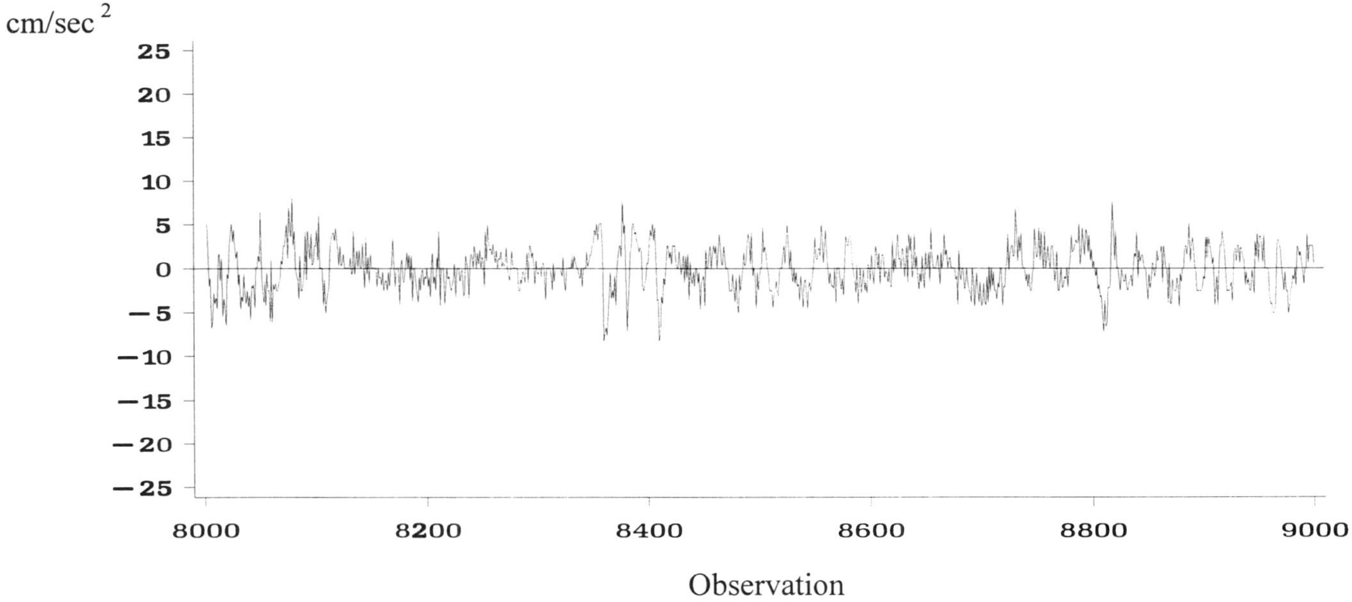

Figure 8-16 Landers, CA Earthquake Change In Acceleration
(Observations 9000-10000   Units: cm/sec^2)

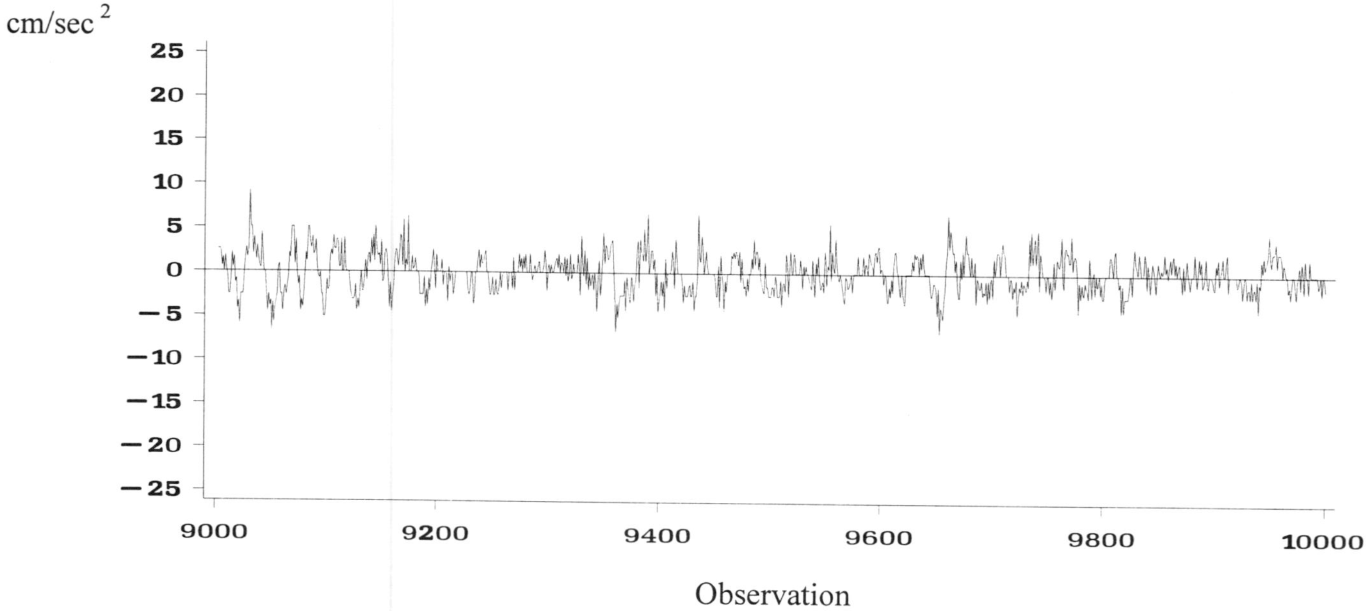

Figure 8-17 Landers, CA Earthquake Change In Acceleration
(Observations 10000-11000 Units: cm/sec ^{2})

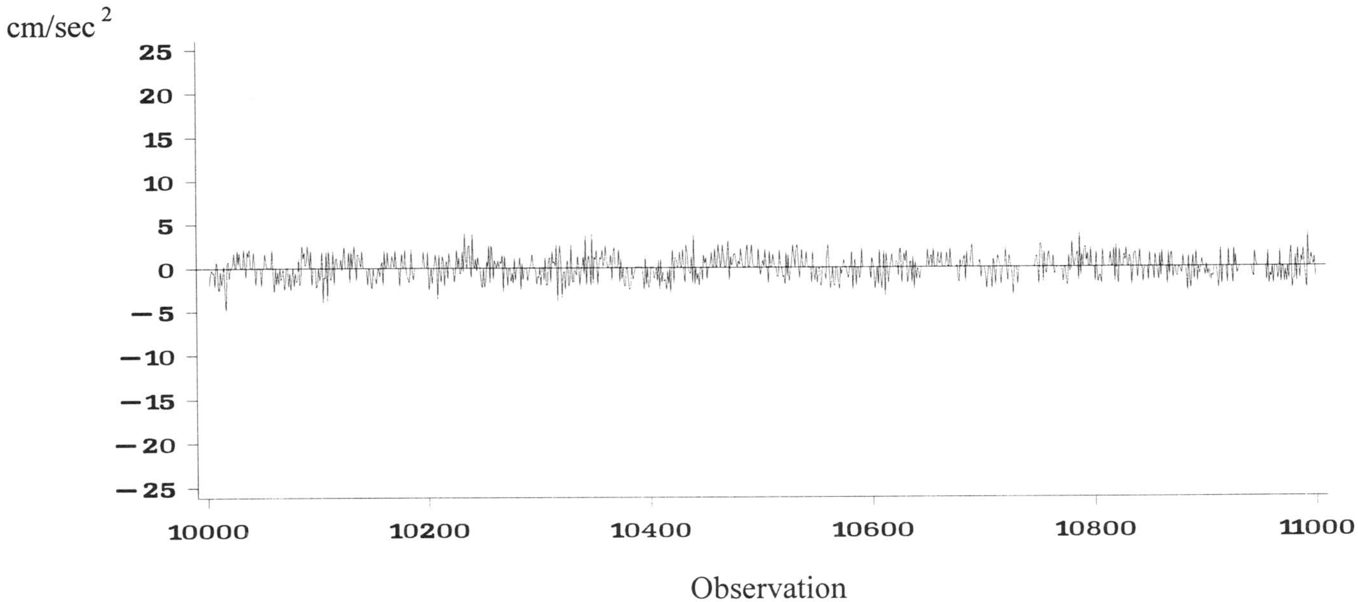

References

Box, G.E.P. and G. M. Jenkins (1976). Time Series Analysis: Forecasting and Control. Holden Day: San Francisco.

Granger, C.W.J. and P. Newbold (1977). *Forecasting Economic Time Series.* Academic Press: New York.

Notes

1. Plots of the data from observation number 1 to 1000 and from observation number 1001 to observation number 8000, 9000, or 10,000 indicate that the variance of the data is substantially smaller outside of this interval.

2. Essentially identical sample autocorrelation results are obtained using the complete sample.

9 Analysis of U.S. Real GNP

The quintessential macroeconomic time series is U.S. real GNP; this is the U. S. Department of Commerce's principal measure of the value of U. S. production of goods and services, corrected for changes in the aggregate price level. The battery of tests analyzed above was applied to the logarithmic growth rate of U.S. real GNP (y_t) over a sample of 163 quarters from the first quarter of 1953 through the third quarter of 1993, or 1953:1 to 1993:3. These data are plotted in Figure 9-1 below; they appear to be reasonably stationary over this time period. The test results themselves are given in Table 9-1.

As expected from results in Ashley and Patterson (1989) on the U.S. Index of Industrial Production and results in Altug, Ashley, and Patterson (1999) and in Potter (1995) on real GNP itself, the null hypothesis of a linear generating mechanism for this time series can be rejected at the 5% level. In fact, this null hypothesis can be rejected at the 2% level using the Tsay test and the Hinich bicovariance test.

The *pattern* of test results for this time series is quite interesting, however. For one thing, we would expect the BDS test to yield the strongest rejections rather than the Tsay and Hinich bicovariance tests, whose power is much lower than that of the BDS test in most of the simulations reported in Chapter 5. Note also that the BDS test rejects the null hypothesis at the 3-5% level of significance for embedding dimension (m) equal to 3 or 4, but does not reject at all for m = 2. This pattern suggests that the power of the BDS test for m=2 is substantially lower than for higher values of m against this particular alternative. In contrast, in the simulations reported in Chapter 5, the power of the BDS test is usually quite similar across the three values of m. Also, it is notable that, since neither the McLeod-Li nor the Engle LM test rejects the null, the nonlinear generating mechanism for real GNP is not inducing significant conditional heteroskedasticity in the series.[1]

Nowadays, real GNP is commonly modeled as a two-state regime switching process of one sort or another. And the Tsay test, which has high relative power against the simple TAR alternative considered in Chapter 5, does reject the null hypothesis at the 2% level for these data. But the pattern of power results in Table 9-1 is quite different from the results in Chapter 5: for data generated from the simple TAR model considered there, the Hinich bicovariance test has quite low power and the BDS test has relatively high power for all three values of the embedding dimension. For data generated from the simple Markov regime switching model considered in Chapter 5, both the Tsay and the Hinich bicovariance tests have low power relative to the BDS

Figure 9-1 Growth Rate in Real GNP From:
1st Quarter 1953 Through 3rd Quarter 1993

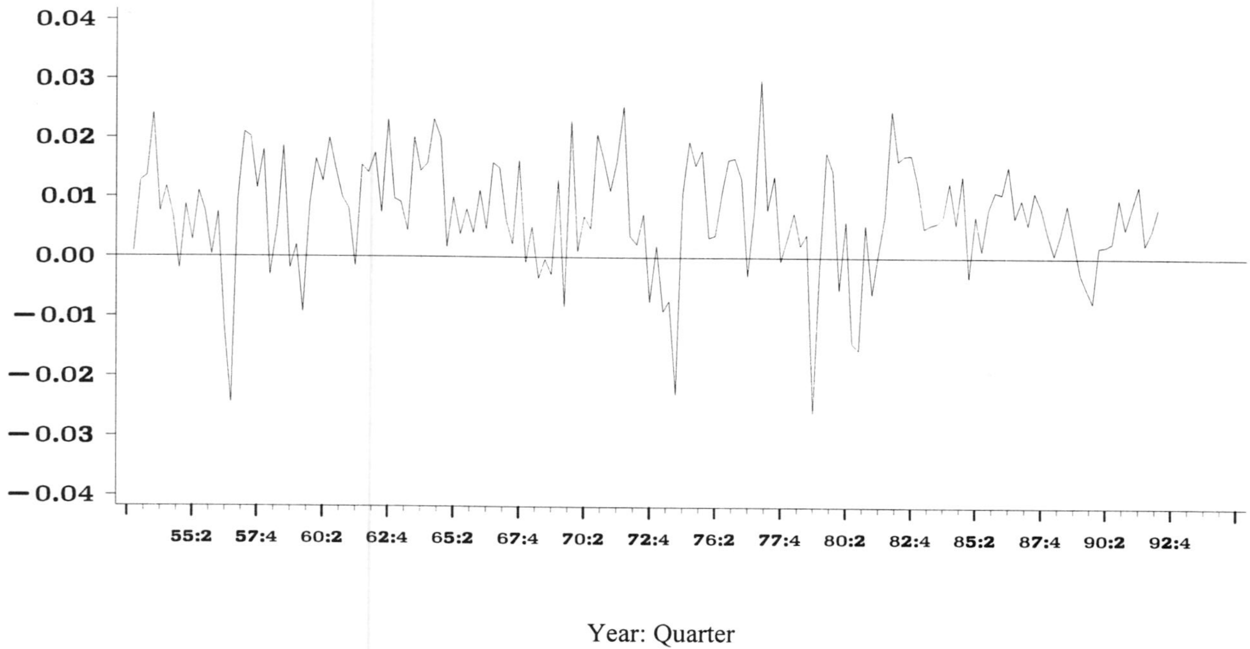

Table 9-1 Significance Levels for Nonlinearity Tests on U.S. Real GNP[a]
Sample Period: 1953:1 to 1993:3

McLeod-Li	Engle LM	BDS			Tsay	Bicov.	Bispectral
$L = 24$	$p = 5$	$m=2$	$m = 3$	$m = 4$	$k = 5$	$\ell = 7$	$M = 21$
.22	.52	.36	**.04**	**.03**	**.02**	**.02**	.33

[a]Test results given in bold are rejections at the 5% level of the null hypothesis of a linear generating mechanism. The significance level for each test was obtained using 1000 bootstrap replications. The parameters L, p, m, ℓ, k, and M are defined in Chapter 2, where each test is discussed.

test and the power of the BDS test again depends little on m. In contrast, using the actual data on real GNP, we see a fairly strong rejection from the Hinich bicovariance test and the odd pattern of BDS results as a function of embedding dimension noted above.

In Ashley and Patterson (1999) we simulate data from several estimated models for real GNP in the literature, a TAR model due to Potter (1995) and several Markov switching models due to Lam (1989), allowing us to estimate the power of each nonlinearity test against these three regime switching alternatives. In essence, we take the pattern of nonlinearity test results given in Table 9-1 to be a new "stylized fact" concerning real GNP and ask whether or not each of these putative models for real GNP can reproduce it.[2]

In fact, none of them can. Except for the Hinich bispectral test, all of the tests appear to have high power to detect the nonlinearity in the data generated from Potter's estimated TAR model; with power this high one would expect the McLeod-Li and Engle LM tests to reject linearity in the actual data if it were generated by a model similar to Potter's TAR, but they do not. In contrast, none of the tests seems particularly effective at detecting the nonlinearity in the data generated using either of the two regime switching models estimated by Lam (1997). With power this low, one would not expect the BDS, Tsay, and Hinich bicovariance tests to reject linearity in the actual data if they were generated by a Markov chain model such as these, but they do reject. Moreover, the BDS test has roughly equal power at each of the three embedding dimensions using data generated from any of the three estimated models. Thus, the pattern of which tests reject linearity using the actual real GNP data conflicts with the pattern of power results obtained using data generated from each of the three estimated models.

These results strongly suggest that the true generating mechanism for U.S. real GNP is not well approximated by either kind of regime switching model. What kind of nonlinear process *does* generate real GNP then? An ARCH or GARCH model seems quite unlikely in view of the failures of the McLeod-Li and Engle LM tests to reject linearity. A quadratic or a cubic model seems possible, or perhaps some form of nonlinear model yet to be proposed.

References

Altug, S., Ashley, R., and Patterson, D. M. (1999) "Are Technology Shocks Nonlinear?"*Macroeconomic Dynamics*, forthcoming.

Ashley, R. and Patterson, D. M. (1989). "Linear Versus Nonlinear Macroeconomies" *International Economic Review 30*, 685-704.

Ashley, R. and Patterson, D.M. (1999). "Nonlinear Model Specification/Diagnostics: Insights From a Battery of Nonlinearity Tests, " Virginia Tech Economics Department Working Paper E99-05, full text available at http://ashleymac.econ.vt.edu/working_papers/E9905.pdf.

Brock, W., J. Lakonishok, and B. LeBaron (1992) "Simple Technical Trading Rules and the Stochastic Properties of Stock Returns" *Journal of Finance XLVII(5)*, 1731-1764.

Potter, S. (1995) "A Nonlinear Approach to U.S. GNP" *Journal of Applied Econometrics 10*, 109-125.

Lam, P. (1997) "A Markov Switching Model of GNP Growth With Duration Dependence" (unpublished manuscript).

Notes

1. Similar results are obtained using the seasonal growth rate of data which has not been seasonally adjusted. There the Tsay test and the Hinich bicovariance test both reject the null of a linear generating mechanism at the 1% level, the BDS test rejects the null at the 2% level (but this time only for $m = 2$), and the other tests do not reject at all at the 5% level.

2. Our approach is similar in spirit to that of Brock, Lakonishok, and LeBaron (1992) in which a putative underlying "null hypothesis" model (random walk, AR(1), GARCH-M, etc.) is fit to daily stock price data (the Dow Jones Industrial Average) and used to bootstrap artificial stock price data. The results of several simple (chartist) trading rules applied to the actual stock price data are compared to those obtained applying the same trading rules to the artificial data generated from each model, allowing Brock, et al. to infer whether each "null" model could have generated the actual data.

10 Dynamic Structure of Macroeconomic Technology Shocks[1]

INTRODUCTION

In the Real Business Cycle approach to macroeconomic theory, the business cycle is taken to be driven by exogenous technological shocks. {E.g., see Kydland and Prescott (1982), Prescott (1986).} Solow (1957) showed that an observable measure of exogenous technological shocks is provided by the difference between the rate of growth of output and the share-weighted growth rates of the inputs to production if there are constant returns to scale, all input factors are fully variable, and there is perfect competition in the product and factor markets.

Thus, letting y_t denote aggregate real output, this observable measure of technological shocks – the so-called "Solow residual" – is given by z_t in

$$y_t = z_t A L_t^{\alpha_t} S_t^{1-\alpha_t}$$

where L_t is aggregate employment and S_t is a measure of the capital services provided by the aggregate capital stock. Thus, the growth rate in the Solow residual is given by

$$\Delta\ln(z_t) = \Delta\ln(y_t) - \alpha_t \Delta\ln(L_t) - (1-\alpha_t)\Delta\ln(S_t)$$

where α_t is labor's share of total income in period t.

DATA AND RESULTS

Real output (y_t) is measured as gross national product (GNP) in 1987 dollars from the National Income and Product Accounts, Table 1.10. Aggregate employment (L_t) is measured as total hours worked, obtained by multiplying CITIBASE variable LHOURS (manhours worked per week for all workers, all industries) by the number of weeks in a quarter. Following Griliches and Jorgenson (1967) and Burnside, Eichenbaum and Rebelo (1995a,b), capital services (S_t) are measured by aggregate electricity usage; this measure allows for time variation in capacity utilization rates.[2]

Aggregate electricity usage is measured by three month average of CITIBASE variable IPCOE, a monthly index of electric utility sales to commercial and other users.[3]

$\Delta\ln(z_t)$, $\Delta\ln(y_t)$,$\Delta\ln(L_t)$, and $\Delta\ln(S_t)$ are plotted versus time in Figures 10-1, 10-2, 10-3, and 10-4 over a 158 quarter sample period running from the second quarter of 1954 to the third quarter of 1993, or 1954:2 to 1993:3.[4] Nonlinearity test results for these four series are given in Table 10-1.

From the first row of Table 10-1 we see that there is no evidence allowing us to reject the null hypothesis of a linear generating mechanism for the Solow residual. Yet the results in the second row, as in Chapter 9 over a slightly longer sample period, indicate that this null hypothesis of a linear generating mechanism **can** be rejected at the 2% level for real output. Evidently, this nonlinearity in the real output series is not due to nonlinearity in the generating mechanism for the exogenous technological shocks: we conclude that the important nonlinearities are in the macroeconomy itself.

The results given in the remaining rows of Table 10-1 further localize the source of the nonlinearity in real output: it is in the markets for labor. The pattern of the test results for the hours worked growth rate series is quite unlike any observed in our simulations – apparently it is the Hinich bicovariance test that has the highest power against the nonlinear mechanism generating this series. This suggests that asymmetries showing up in the third moments are an important feature of these data and yields a rather specific stylized fact that any putative generating mechanism for hours worked must be able to reproduce.

Figure 10-1 Growth Rate in Solow Residual
Sample Period: 1954:2 to 1993:3

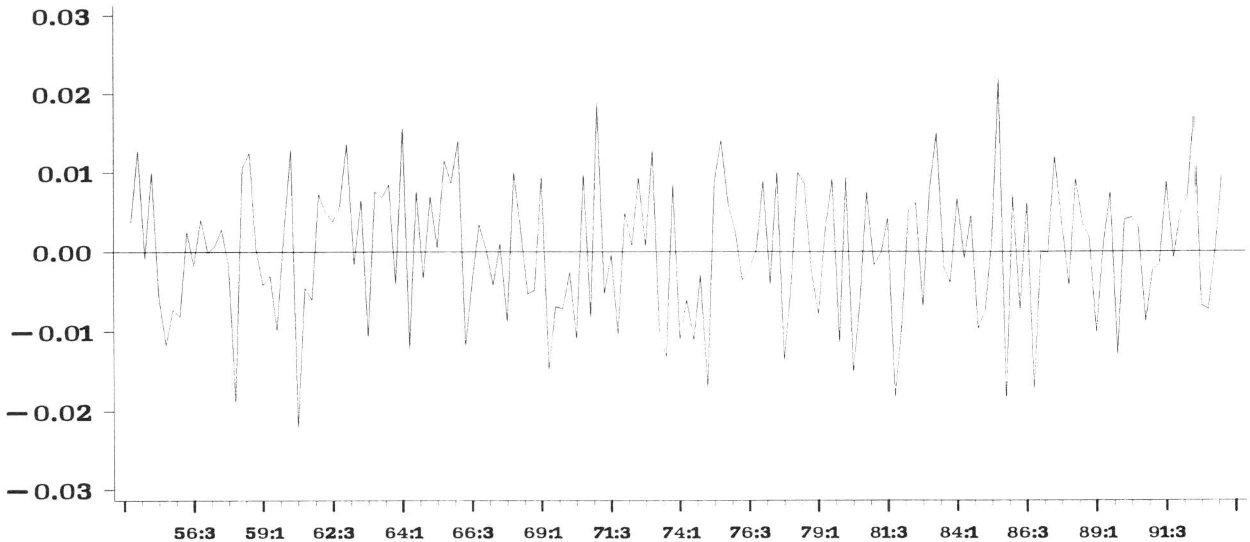

Year: Quarter

Figure 10-2 Growth Rate in Capital Services (Electricity Usage)
Sample Period: 1954:2 to 1993:3

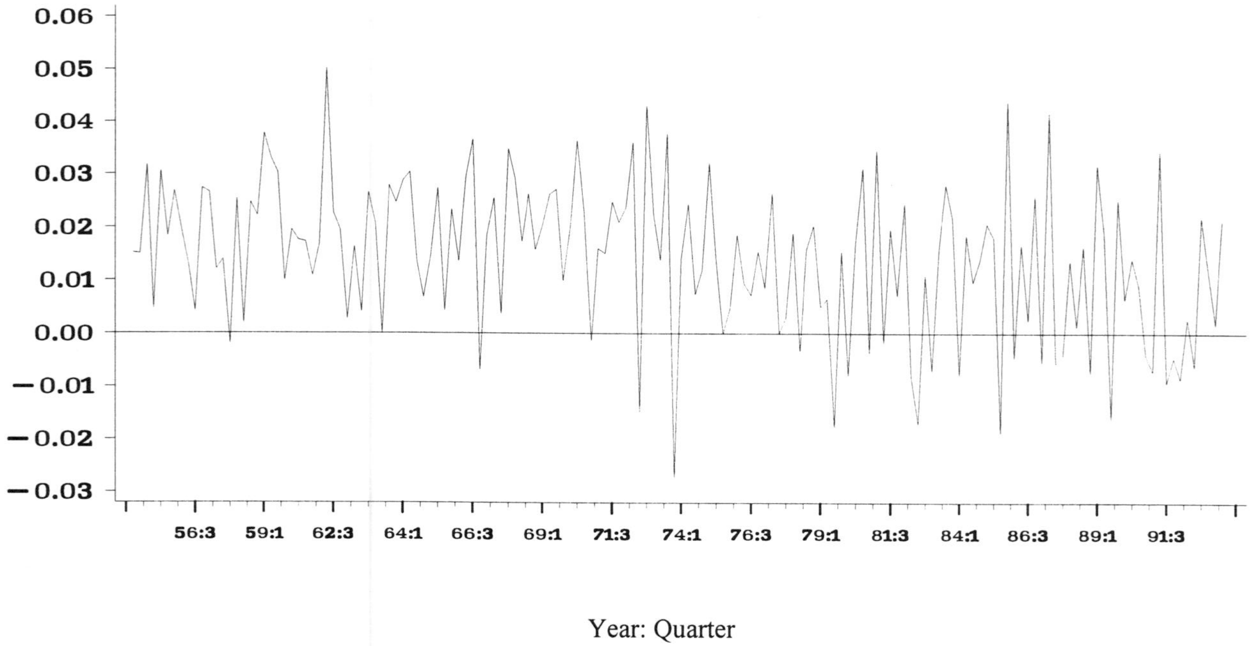

Figure 10-3 Growth Rate in Hours Worked
Sample Period: 1954:2 to 1993:3

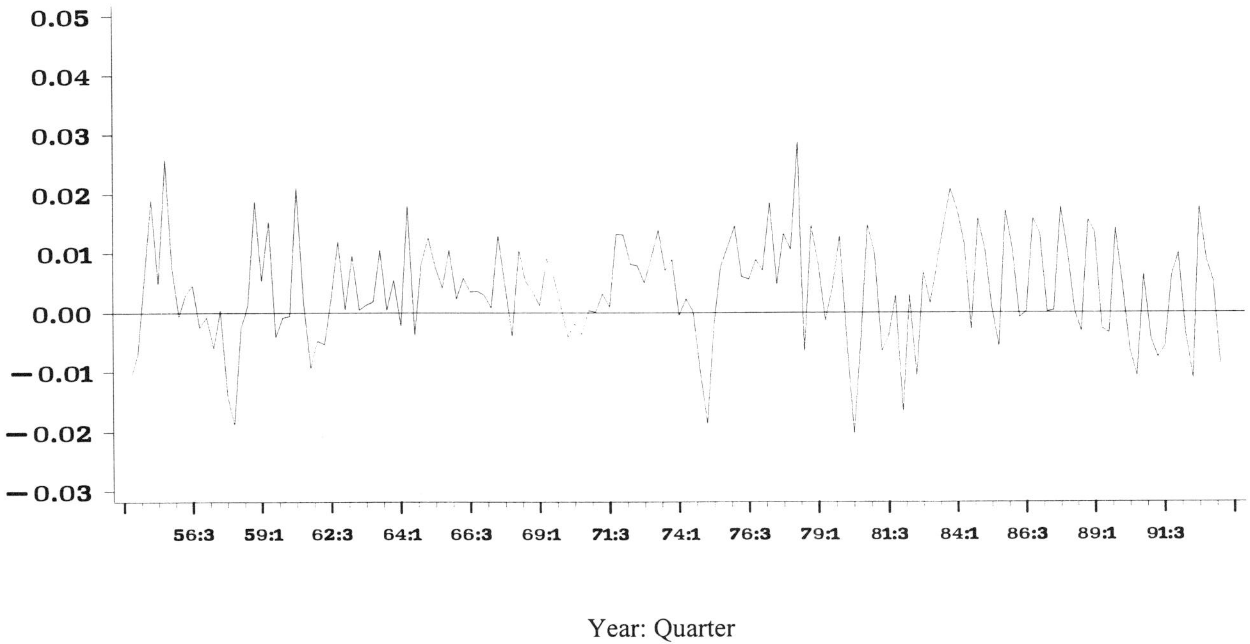

Figure 10-4 Growth Rate in Real GNP
Sample Period: 1954:2 to 1993:3

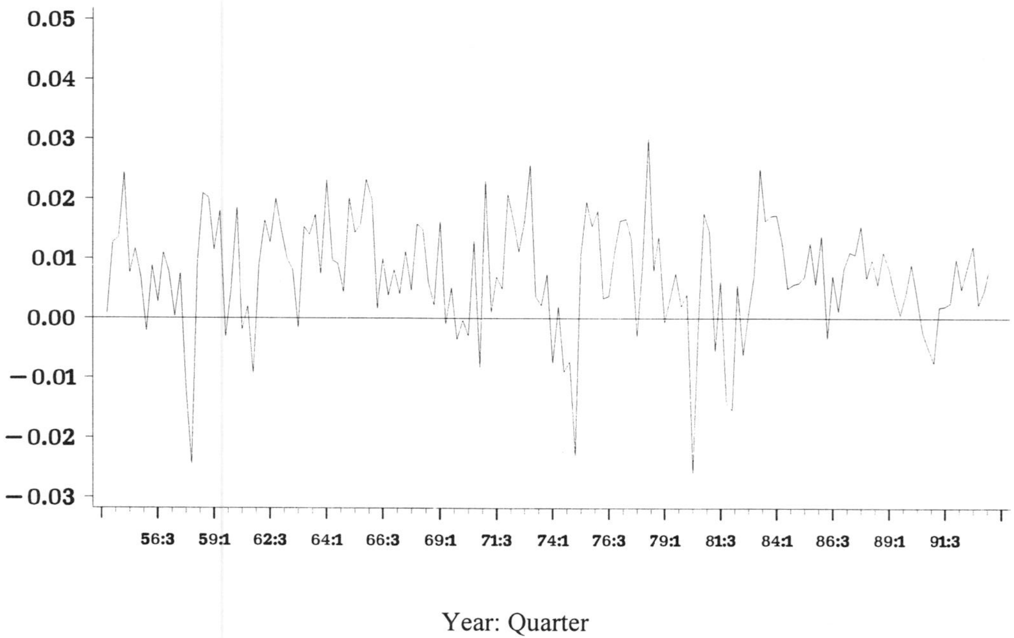

Table 10-1 Nonlinearity Test Results on Solow Residual Growth Rates and Related Series[a]
Sample Period: 1954:2 to 1993:3

	McLeod-Li	Engle LM	BDS			Tsay	Bicovariance	Bispectral
	$L = 24$	$p = 5$	$m = 2$	$m = 3$	$m = 4$	$k = 5$	$\ell = 7$	$M = 21$
Solow Residual $\Delta\ln(z_t)$	.58	.34	.53	.54	.68	.80	.32	.29
Real Output $\Delta\ln(y_t)$	.22	.53	.36	**.04**	**.03**	**.02**	**.02**	.33
Labor Input $\Delta\ln(L_t)$	.16	.71	.18	.08	.07	.14	**.03**	.93
Capital Input $\Delta\ln(S_t)$	.48	.18	.09	.11	.27	.93	.82	.63

[a]All figures quoted are based on 250 generated samples. The 5% critical region for each test was obtained using 1000 bootstrap replications. The parameters L, p, m, ℓ, k, and M are defined in Chapter 2, where each test is discussed.

References

Altug, S., R. Ashley, R., and D. M. Patterson (1999) "Are Technology Shocks Nonlinear?" *Macroeconomic Dynamics*, forthcoming.

Burnside, C., M. Eichenbaum, and S. Rebelo (1993). "Labor hoarding and the business cycle."*Journal of Political Economy 101*, 245--273.

Burnside, C., M. Eichenbaum, and S. Rebelo (1995). "Capital utilization and returns to scale.*NBER Macroeconomics Annual* Cambridge, MA: MIT Press.

Griliches, Z. and D. Jorgenson (1967). "The explanation of productivity change." *Review of Economic Studies 34*, 249--282.

Kydland, F. and E. Prescott (1982). "Time-to-build and aggregate fluctuations. *Econometrica 50*, 1345--1370.

Prescott, E. (1986). "Theory ahead of business cycle measurement."*Carnegie-Rochester Conference Series on Public Policy 25*, 11-44.

Solow, R. (1957). "Technical change and the aggregate production function." *Review of Economics and Statistics 39*, 312--320.

Notes

1.Material in this chapter is drawn from Altug, Ashley and Patterson (1999).

2.Altug, Ashley and Patterson (1999) also use a measure of S_t based on estimates of the aggregate net capital stock. This measure yields a growth rate series $\{\Delta ln(S_t)\}$which is still I(1), however, so it is not considered here.

3.The growth rate in this series $\{\Delta ln(S_t)\}$ has an outlier (t = -3.6) in the first quarter of 1974 due to the disruption in crude oil supplies from the Middle East in the previous quarter; this was eliminated by setting this observation equal to the sample mean. This adjustment did not materially affect the test results on the Solow residual, but did affect the results on $\Delta ln(S_t)$ itself.

4.This is the same sample period used in Altug, Ashley and Patterson (1999). This sample begins a few quarters later than that of Chapter 9 due to unavailability of IPCOE data in the previous periods.

11 Climatological Data

INTRODUCTION

Over the last 3 million years or so "ice ages" – brief periods of colder weather allowing large sheets of ice to develop – have been the dominant long term feature of Earth's climate.

Climatologists use $\delta^{18}O$, the oxygen isotope ratio in ocean sediments (expressed as the percentage of total oxygen in the form of ^{18}O) as a proxy for the total volume of water tied up in ice sheets. Figure 11-1 plots $\delta^{18}O_t$ versus time for a typical set of sediment cores obtained by the Deep Sea Drilling Project (DSDP) at site number 607. DSDP site 607 was drilled from a depth of 3427 meters below sea level at latitude 41°00' N and longitude 32°58' W, placing it in the Atlantic ocean roughly halfway between New York and Madrid. Here we follow the climatologists' convention and plot $\delta^{18}O_t$ versus time in kyrs "BP" – or, "before the present." Thus, time is actually increasing from right to left in all figures in this Chapter.

DATA

Inspection of this time plot indicates that there are three separate, roughly stationary regimes in this time series. The most recent period (700 - 0 kyrs BP, more or less, separately plotted in Figure 11-2) is known to be dominated by a "100 kyr" cycle, corresponding to the last seven ice ages. Like business cycles, the lengths of these cycles are not constant, varying from 84 kyr to 120 kyr over the period. The variance of $\delta^{18}O_t$ is smaller and the apparent period of the cycle shorter over the previous 2000 kyr or so, forming the second regime. And the remaining data, prior to 2700 kyr BP seems different still; it will not be further considered here. To compare with the 700 - 0 kyr BP record plotted in Figure 11-2, we have also selected a similarly stationary 700 kyr sample, running from 1700 - 1000 kyr BP, from the middle regime. These data are plotted in Figure 11-3.

Both of these time series seem quite smooth. Thus, our first task is to determine whether they are I(0) or I(1) – i.e., whether they require differencing in order to be covariance stationary. Following Box and Jenkins (1976) and Granger and Newbold (1977), we first compute the sample autocorrelation function for each series; these are given in Table 11-1 and plotted in Figures 11-4 and 11-5 below.

The sample autocorrelations for both $\delta^{18}O_t$ series decay to zero much more promptly than one would expect for I(1) data[1]. But the autocorrelation at lag 1, corresponding to an interval of 2 kyr, is too close to one for this argument to be compelling. Consequently, we also used the augmented Dickey-Fuller test (Dickey

Figure 11-1 DSDP Site 607: Oxygen 18 versus Time
(All Data)

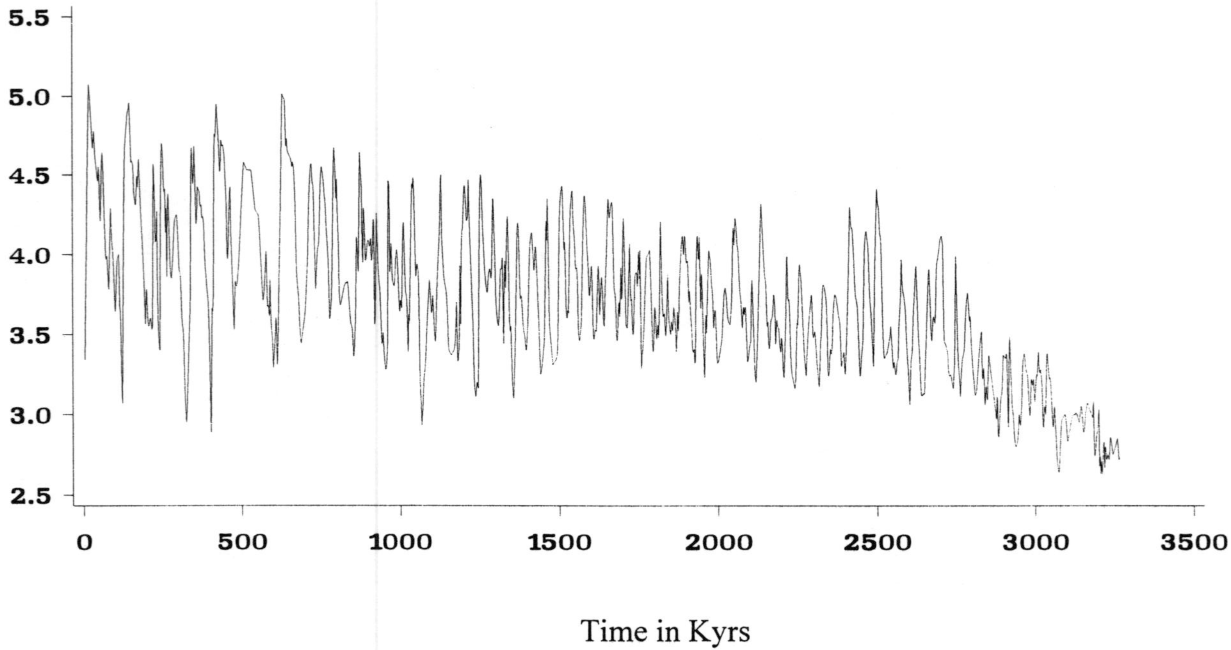

Figure 11-2 DSDP Site 607: Oxygen 18 versus Time
(0 - 800 Kyrs)

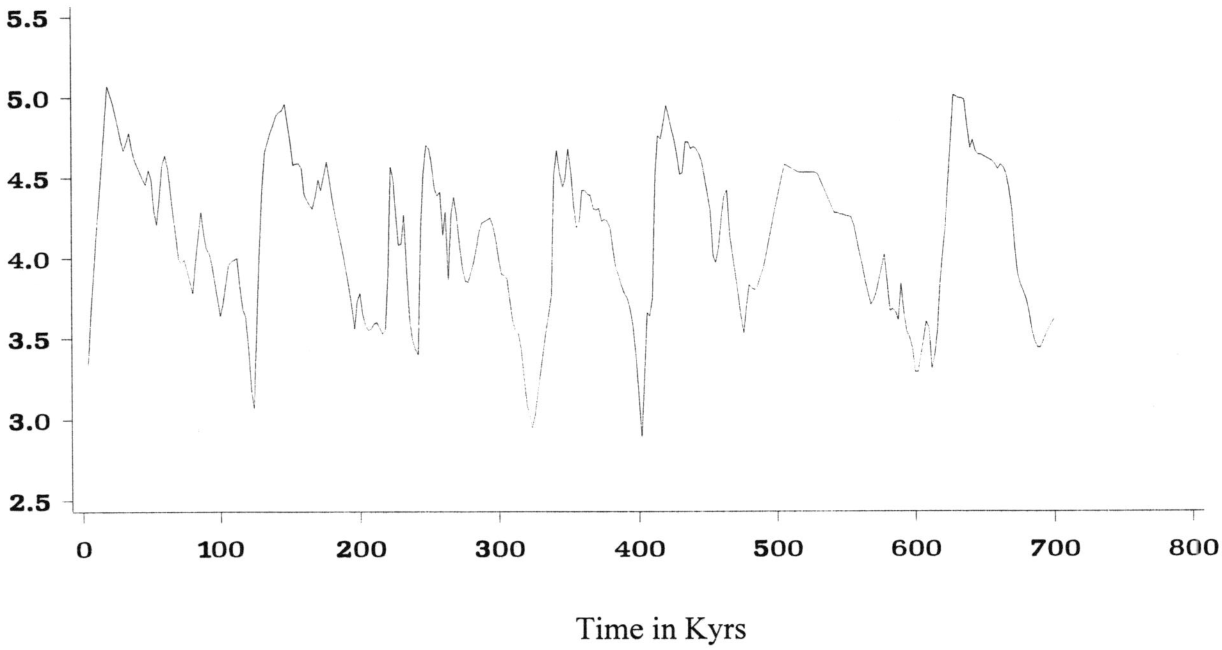

Figure 11-3 DSDP Site 607: Oxygen 18 versus Time
(1000 - 1700 Kyrs)

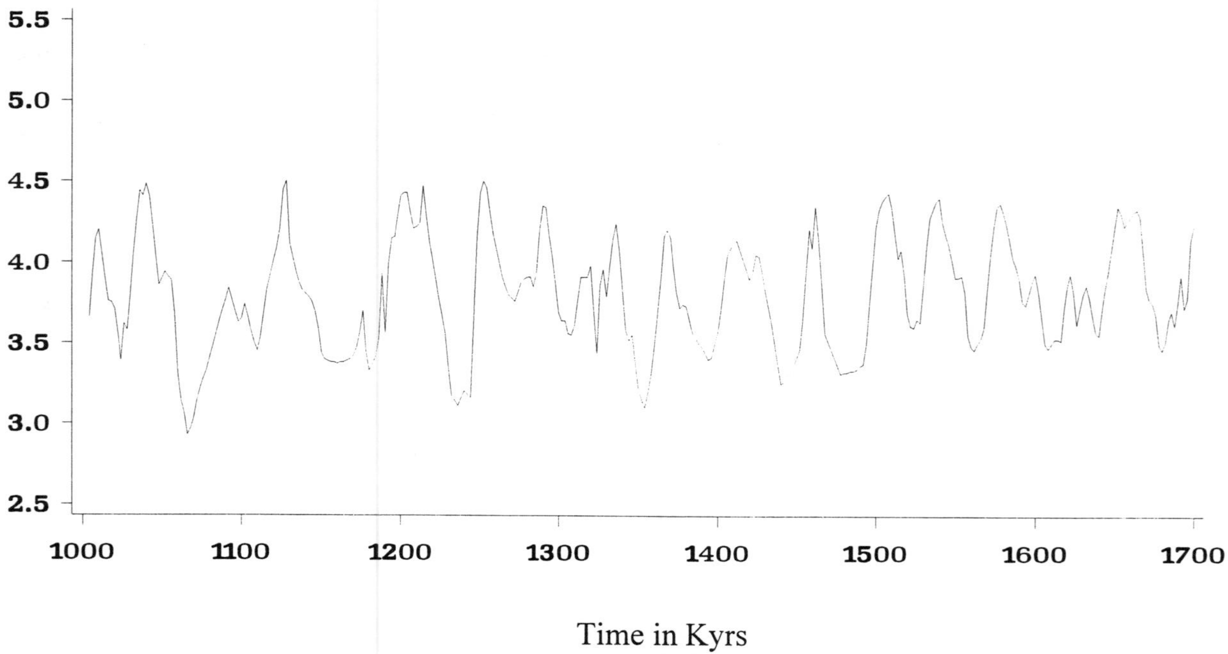

Table 11-1
Sample Autocorrelations $\delta^{18}O_t$ data[a]

lag	700 - 0 kyrs	1700 - 1000 kyrs
1	.942	.922
2	.836	.776
3	.721	.604
4	.606	.416
5	.491	.232
6	.381	.067
7	.282	-.077
8	.199	-.189
9	.136	-.264
10	.084	-.307
11	.041	-.323
12	-.011	-.319
13	-.069	-.297
14	-.126	-.257
15	-.181	-.199

[a]Data is available at 2 kyr intervals, so N = 350 and the lags are in 2 kyr increments. The sampling errors in these estimates of the population autocorrelations are substantially correlated because of the strong low-order autocorrelation in the series itself. See Bartlett (1946).

Figure 11-4 Sample Autocorrelations $\delta^{18}O_t$ data
(700 - 0 kyrs)

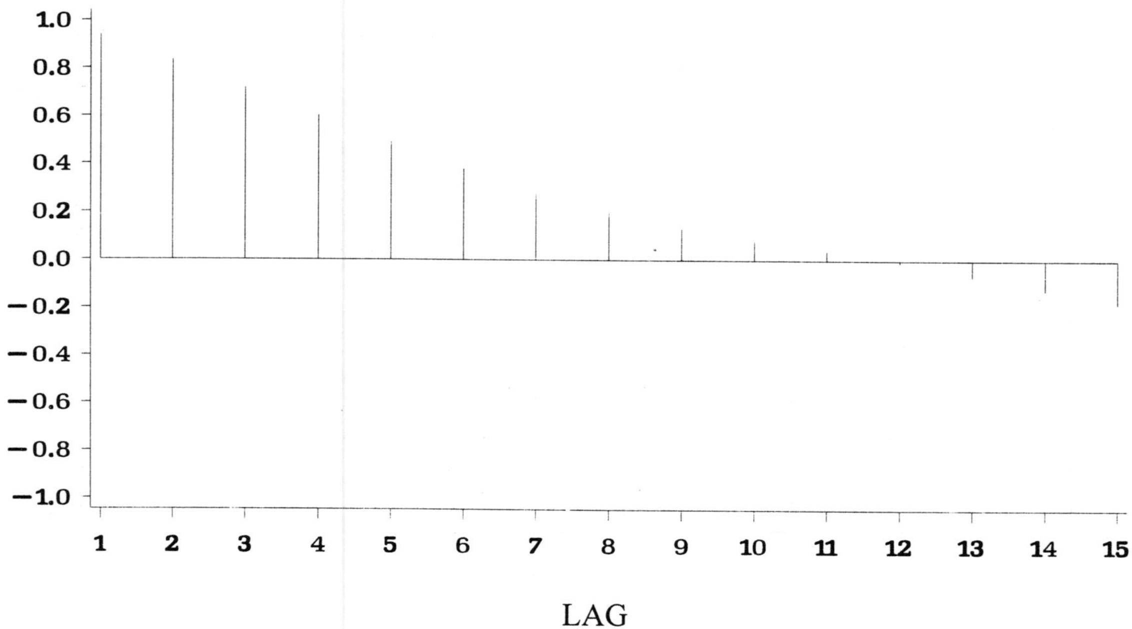

Figure 11-5 Sample Autocorrelations $\delta^{18}O_t$ data
(1700 - 1000 kyrs)

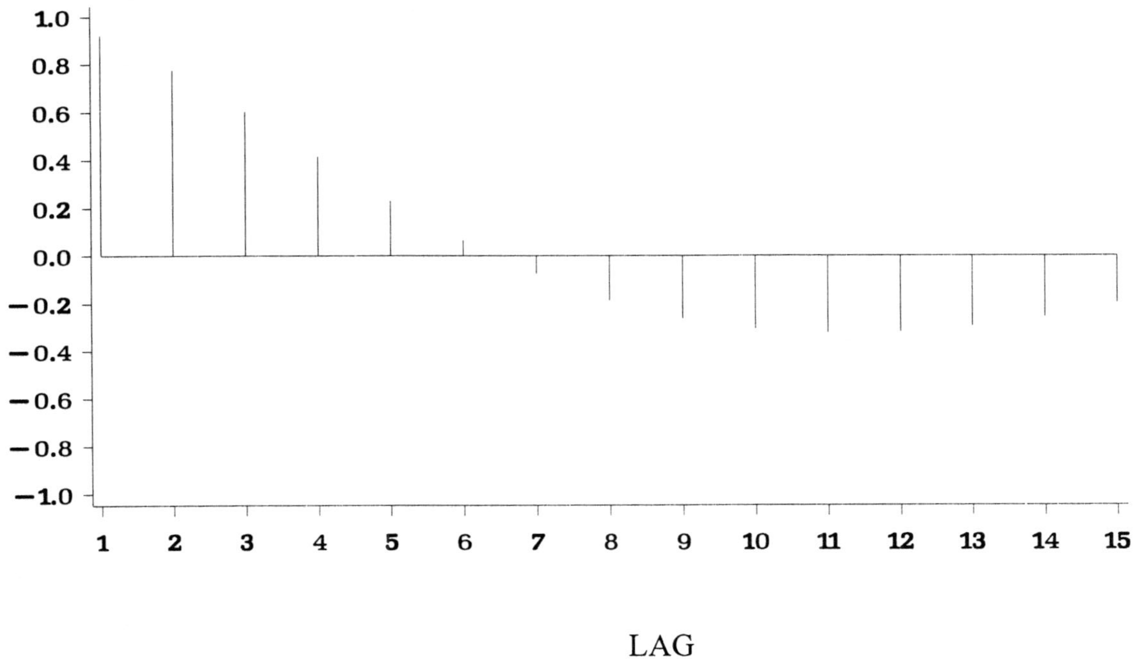

and Fuller (1981)) to formally test each of these times series for a unit root.

The augmented Dickey-Fuller test involves regressing the change in the series ($\Delta\, \delta^{18}O_t = \delta^{18}O_t - \delta^{18}O_{t-1}$) against the lagged level of the series ($\delta^{18}O_{t-1}$) and however many lags in $\Delta\, \delta^{18}O_t$ are needed. Here, only the first lag in each differenced series entered with a statistically significant estimated coefficient, yielding the two estimated regression equations:

700 - 0 kyrs:

$$\Delta\left\{\delta^{18}O_t\right\} = \underset{(6.35)}{.433} - \underset{(-6.37)}{.114}\, \delta^{18}O_{t-1} + \underset{(11.17)}{.510}\, \Delta\left\{\delta^{18}O_{t-1}\right\} + e_t$$

1700 - 1000 kyrs:

$$\Delta\left\{\delta^{18}O_t\right\} = \underset{(5.01)}{.320} - \underset{(-5.04)}{.077}\, \delta^{18}O_{t-1} + \underset{(10.22)}{.484}\, \Delta\left\{\delta^{18}O_{t-1}\right\} + n_t$$

Since the magnitude of the estimated t ratio on the lagged level series in each of these regression equations substantially exceeds 3.48, the 1% critical value of the Dickey-Fuller test for N = 350, we reject the null hypothesis of a unit root in both of these time series. Therefore, we continue our analysis using the level (rather than the first difference) of each series.

RESULTS

We then placed both data sets in actual time order and applied the battery of nonlinearity tests to them[2]. The results are given in Table 11-2 below. Overall, these results strikingly confirm the (well known) fact that, as Raymo (1997) observes,

> climate cycles (in particular global ice volume) are highly nonlinear ... the cycles are "saw tooth" shaped, with long periods of ice growth/cooling followed by rapid deglaciation ...

But it is interesting to observe that the test rejections of a linear generating mechanism seem equally strong for both periods, despite the fact that the "saw tooth" pattern is substantially more prevalent in the more recent period and the fact that the huge "100 kyr" ice sheets are only able to form in this latter period.

Table 11-2
Significance Levels for Nonlinearity Tests on DSPD Site 607 $\delta^{18}O$ data[a]

	McLeod-Li	Engle LM	BDS			Tsay	Bicov.	Bispectral
Period (kyr)	L = 24	p = 5	m=2	m = 3	m = 4	k = 5	ℓ = 10	M = 34
700 - 0	**.00**	**.01**	**.00**	**.00**	**.00**	**.03**	**.00**	.33
1700 - 1000	**.00**	**.00**	**.00**	**.00**	**.00**	.24	**.00**	**.04**

[a]Test results given in bold are rejections at the 5% level of the null hypothesis of a linear generating mechanism. The 5% critical region for each test was obtained using 1000 bootstrap replications. The parameters L, p, m, ℓ, k, and M are defined in Chapter 2, where each test is discussed.

Muller and MacDonald (1997) state that "a climate model that depends on forced oscillation has a spectrum that reflects the spectrum of the driving force" and go on to propose a particular astronomically forced model based in large part on the positions of narrow spectral peaks in $\delta^{18}O_t$ time series such as those considered here.[3] However, the strong nonlinearity we observe in these data indicates that substantial caution must be exercised in interpreting the estimated spectrum of $\delta^{18}O_t$ time series from ocean sediment data – the spectrum, after all, is based on the autocovariance function, which completely characterizes only *linearly* generated data.

While it is more typical for a nonlinear process to smear sharp spectral features in a forcing input or to create spurious side bands around them, it is also possible for nonlinear systems to create sharp features in the power spectrum of the output where there were none in the forcing inputs. Thus, Muller and MacDonald's model may be correct, but our finding of strong nonlinearity in these data leaves the spectral evidence in favor of their model less than compelling.

Finally, we note that the *pattern* of the test results is quite different for the two periods: the Tsay test fails to reject for the earlier period and the bispectral test fails to reject for the later period. This result is consistent with the observation that huge ice sheets formed in the second but not in the first period. This pair of patterns constitutes a stylized fact which any satisfactory quantitative theory of this phenomenon must be able to duplicate.

REFERENCES

Bartlett, M..S. (1946) "On the Theoretical Specification of Sampling Properties of Autocorrelated Time Series." *Journal of the Royal Statistical Society B 8*, pp. 27-41.

Box, G.E.P. and G. M. Jenkins (1976). Time Series Analysis: Forecasting and Control. Holden Day: San Francisco.

Dickey, D. and W. Fuller (1981). "Likelihood Ratio Tests for Autoregressive Time Series with a Unit Root." *Econometrica 49*, pp. 1057-72.

Granger, C.W.J. and P. Newbold (1977). *Forecasting Economic Time Series.* Academic Press: New York.

Muller, R. A. and G. J. MacDonald (1997) "Glacial Cycles and Astronomical Forcing." *Science 277*, pp. 215-218.

Raymo, M. (1997). "The Timing of Major Climate Terminations." *Paleoceanography 12(4)*, pp. 577-85.

Notes

1.The sampling errors in these estimates of the population autocorrelations are substantially correlated because of the strong low-order autocorrelation in the series itself. See Bartlett (1946).

2.Nonlinearity tests, like the processes they test for, are in general not time reversible.

3.The spectral peaks they observe (e.g., in their Figure 1a) are indeed fairly narrow: in fact, they are considerably narrower than one might expect, given the observed variations in the lengths of the cycles. But this appearance of a spectrum dominated by a single narrow peak is largely due to the fact that Muller and MacDonald plot the spectrum itself versus frequency: more commonly, the power spectrum – the logarithm of the spectrum – is plotted against frequency. Indeed, when the power spectrum for the DSDP site 607 data from 2500 - 1500 kyr is plotted against frequency rather than the spectrum itself, the huge peak Muller and MacDonald observe with period 41 kyr appears only marginally larger than other peaks at periods of 22 and 16 kyrs.

INDEX

Dynamic Modeling and Econometrics in Economics and Finance

1. P.Rothman:
   Nonlinear Time Series Analysis of Economic and Financial Data. 1999 — ISBN 0 7923-8379-6
2. D.Patterson,R.Ashley:
   A Nonlinear Time Series Workshop: A Toolkit For Detecting and Identifying Nonlinear Serial Dependence. 1999 — ISBN 0-7923-8674-4